Efficiency and Optimization of Buildings Energy Consumption: Volume II

Efficiency and Optimization of Buildings Energy Consumption: Volume II

Editor

José A. Orosa

MDPI • Basel • Beijing • Wuhan • Barcelona • Belgrade • Manchester • Tokyo • Cluj • Tianjin

Editor
José A. Orosa
Navigation Science and
Marine Engineering
University of A Coruña
A Coruña
Spain

Editorial Office
MDPI
St. Alban-Anlage 66
4052 Basel, Switzerland

This is a reprint of articles from the Special Issue published online in the open access journal *Applied Sciences* (ISSN 2076-3417) (available at: www.mdpi.com/journal/applsci/special_issues/buildings_energy_consumption_volume_II).

For citation purposes, cite each article independently as indicated on the article page online and as indicated below:

LastName, A.A.; LastName, B.B.; LastName, C.C. Article Title. *Journal Name* **Year**, *Volume Number*, Page Range.

ISBN 978-3-0365-6508-8 (Hbk)
ISBN 978-3-0365-6507-1 (PDF)

Contents

About the Editor

José A. Orosa

José A. Orosa currently works at the University of A Coruña, E.T.S Náutica y Máquinas. He was identified as the second most relevant researcher over the last ten years (2006-2016) in Sustainability and Energy efficiency by the research work titled "Worldwide Research on Energy Efficiency and Sustainability in Public Buildings (Sustainability 2017, 9, 1294; doi:10.3390/su9081294)". His current project is "Energy and Resources Efficiency in Building Codes in West Africa".

Preface to "Efficiency and Optimization of Buildings Energy Consumption: Volume II"

This reprint gives an up-to-date overview of new technologies based on Machine Learning (ML) and Internet of Things (IoT) procedures to improve the mathematical approach of algorithms that allow control systems to be improved with the aim of reducing housing sector energy consumption.

To achieve the objective of reducing buildings'energy consumption, some papers aim to improve building constructive characteristics and materials, but always within the realistic economical and health limitations, and others look for other energy sources implemented with different control system algorithms. The key is to define a correct algorithm based on correct variables (such as weather conditions) and, at other times, to employ a more adequate machine learning method.

In this reprint, different options for the quantification of energy saving are shown and these results will be of interest to engineers and researchers interested in energy saving and its optimization in buildings.

José A. Orosa
Editor

Editorial

Efficiency and Optimization of Buildings Energy Consumption Volume II

José A. Orosa

Department of Navigation Science and Marine Engineering, Universidade da Coruña, Paseo de Ronda, 51, 15011 A Coruña, Spain; jose.antonio.orosa@udc.es

1. Introduction

This issue, as a continuation of a previous Special Issue on "Efficiency and Optimization of Buildings Energy Consumption," gives an up-to-date overview of new technologies based on Machine Learning (ML) and Internet of Things (IoT) procedures to improve the mathematical approach of algorithms that allow control systems to be improved with the aim of reducing housing sector energy consumption.

2. Energy Optimization Procedures

To achieve the objective of reducing buildings' energy consumption, some papers aim to improve building constructive characteristics and materials [1], but always within the realistic economical and health limitations [2,3], and others look for other energy sources implemented with different control system algorithms [4,5]. To improve this, construction materials showed an expected reduction of 30% and implementation of a Passivhaus resulted in a reduction of 85% in heating demand [2]. Finally, the control of daylighting accounted for the highest energy savings of 14% [4].

In consequence, the key is to define a correct algorithm based on correct variables (such as weather conditions [6]) and, at other times, to employ a more adequate machine learning method. In this sense, a Support Vector Machine [7] was shown to be adequate for global solar prediction, but short memory neural networks were preferred for the prediction of other classical variables related to energy consumption in buildings, such as thermal inertia and input time lag [8]. More examples of machine learning methods were employed to predict [9] and to define building parameters like the Heat Loss Coefficient [8], reaching a maximum error of 6%. Results showed that weather data control systems may reach 23% [6].

3. Future Tasks

As a pending research task, a smart grid optimized by artificial intelligence and employing Internet of Things technology let a multilayer feed-forward artificial neural network improve the previous results in real case studies [10].

Conflicts of Interest: The authors declare no conflict of interest.

Citation: Orosa, J.A. Efficiency and Optimization of Buildings Energy Consumption Volume II. *Appl. Sci.* **2023**, *13*, 361. https://doi.org/10.3390/app13010361

Received: 26 December 2022
Accepted: 26 December 2022
Published: 27 December 2022

References

1. Borbon-Almada, A.C.; Lucero-Alvarez, J.; Rodriguez-Muñoz, N.A.; Ramirez-Celaya, M.; Castro-Brockman, S.; Sau-Soto, N.; Najera-Trejo, M. Design and Application of Cellular Concrete on a Mexican Residential Building and Its Influence on Energy Savings in Hot Climates: Projections to 2050. *Appl. Sci.* **2020**, *10*, 8225. [CrossRef]
2. Martinez-Soto, A.; Saldias-Lagos, Y.; Marincioni, V.; Nix, E. Affordable, Energy-Efficient Housing Design for Chile: Achieving Passivhaus Standard with the Chilean State Housing Subsidy. *Appl. Sci.* **2020**, *10*, 7390. [CrossRef]
3. Orosa, J.A.; Nematchoua, M.K.; Reiter, S. Air Changes for Healthy Indoor Ambiences under Pandemic Conditions and Its Energetic Implications: A Galician Case Study. *Appl. Sci.* **2020**, *10*, 7169. [CrossRef]
4. Litardo, J.; Palme, M.; Hidalgo-León, R.; Amoroso, F.; Soriano, G. Energy Saving Strategies and On-Site Power Generation in a University Building from a Tropical Climate. *Appl. Sci.* **2021**, *11*, 542. [CrossRef]
5. Weng, L.; Zhang, X.; Qian, J.; Xia, M.; Xu, Y.; Wang, K. Non-Intrusive Load Disaggregation Based on a Multi-Scale Attention Residual Network. *Appl. Sci.* **2020**, *10*, 9132. [CrossRef]
6. Cieśliński, K.; Tabor, S.; Szul, T. Evaluation of Energy Efficiency in Thermally Improved Residential Buildings, with a Weather Controlled Central Heating System. A Case Study in Poland. *Appl. Sci.* **2020**, *10*, 8430. [CrossRef]
7. Álvarez-Alvarado, J.M.; Ríos-Moreno, J.G.; Obregón-Biosca, S.A.; Ronquillo-Lomelí, G.; Ventura-Ramos, J.E.; Trejo-Perea, M. Hybrid Techniques to Predict Solar Radiation Using Support Vector Machine and Search Optimization Algorithms: A Review. *Appl. Sci.* **2021**, *11*, 1044. [CrossRef]
8. Martínez-Comesaña, M.; Febrero-Garrido, L.; Granada-Álvarez, E.; Martínez-Torres, J.; Martínez-Mariño, S. Heat Loss Coefficient Estimation Applied to Existing Buildings through Machine Learning Models. *Appl. Sci.* **2020**, *10*, 8968. [CrossRef]
9. Comesaña, M.M.; Febrero-Garrido, L.; Troncoso-Pastoriza, F.; Martínez-Torres, J. Prediction of Building's Thermal Performance Using LSTM and MLP Neural Networks. *Appl. Sci.* **2020**, *10*, 7439. [CrossRef]
10. Hu, Y.-C.; Lin, Y.-H.; Lin, C.-H. Artificial Intelligence, Accelerated in Parallel Computing and Applied to Nonintrusive Appliance Load Monitoring for Residential Demand-Side Management in a Smart Grid: A Comparative Study. *Appl. Sci.* **2020**, *10*, 8114. [CrossRef]

Article

Design and Application of Cellular Concrete on a Mexican Residential Building and Its Influence on Energy Savings in Hot Climates: Projections to 2050

Ana C. Borbon-Almada [1], Jorge Lucero-Alvarez [2], Norma A. Rodriguez-Muñoz [3,*], Manuel Ramirez-Celaya [1], Samuel Castro-Brockman [1], Nicolas Sau-Soto [1] and Mario Najera-Trejo [4]

1 Departamento de Ingeniería Civil y Minas, Universidad de Sonora, Blvd. Luis Encinas y Rosales, Centro, Hermosillo 83000, Mexico; ana.borbon@unison.mx (A.C.B.-A.); manuel.ramirez@unison.mx (M.R.-C.); samuel.castro@unison.mx (S.C.-B.); nicolas.sau@unison.mx (N.S.-S.)
2 Facultad de Zootecnia y Ecología, Universidad Autónoma de Chihuahua, Periférico Francisco R. Almada, km 1, Chihuahua 31453, Mexico; jluceroa@uach.mx
3 Cátedras CONACYT, Centro de Investigación en Materiales Avanzados, Calle CIMAV 110, Ejido Arroyo Seco, Durango 34147, Mexico
4 Departamento de Ingeniería Sustentable, Centro de Investigación en Materiales Avanzados, Calle CIMAV 110, Ejido Arroyo Seco, Durango 34147, Mexico; mario.najera@cimav.edu.mx
* Correspondence: norma.rodriguez@cimav.edu.mx; Tel.: +52-614-439-4898

Received: 2 November 2020; Accepted: 16 November 2020; Published: 20 November 2020

Featured Application: A cellular concrete design for building walls to promote the reduction of heat transfer on the envelope and the energy-related CO_2 emissions without compromising the mechanical performance of the material.

Abstract: The thermal performance of economical housing located in hot climates remains a pending subject, especially in emerging economies. A cellular concrete mixture was designed, considering its thermophysical properties, to apply the new material into building envelopes. The proposed materials have low density and thermal conductivity to be used as a nonstructural lightweight construction element. From the design stage, a series of wall systems based on cellular concrete was proposed. Whereas in the second phase, the materials were analyzed to obtain the potential energy savings using dynamic simulations. It is foreseen that the energy consumption in buildings located in these climates will continue to increase critically due to the temperature increase associated with climate change. The temperatures predicted mean vote (PMV), electric energy consumption, and CO_2 emissions were calculated for three IPCC scenarios. These results will help to identify the impact of climate change on the energy use of the houses built under these weather conditions. The results show that if the conventional concrete blocks continue to be used, the air conditioning energy requirements will increase to 49% for 2030 and 61% by 2050. The proposed cellular concrete could reduce energy consumption between 15% and 28%, and these saving rates would remain in the future. The results indicate that it is necessary to drive the adoption of lightweight materials, so the impact of energy use on climate change can be reduced.

Keywords: cellular concrete; lightweight materials; thermal conductivity; electricity; dynamic simulation; housing; climate change

---

1. Introduction

The climate change effects over the different areas of life are considered one of the most critical and urgent aspects to attend, especially in the urban areas related to the use and consumption of material goods [1–3]. However, it is known that the Mexican industry builds and uses materials that are not in compliance with the precept of habitability and keeping the buildings within the thermal comfort zone, especially in hot climates [4,5]. The situation means that in order to reach thermal comfort, heating, ventilation, and air conditioning systems (HVAC) are intensively used, generating high electric energy demands affecting the seasonal electricity production and damaging the environment due to the intensive use of fossil fuels [6,7].

Nowadays, construction materials in Mexico present low capacity to prevent heat transfer into the buildings' interior. Most of the materials used in urban construction in Mexico are manufactured from mixtures with cementitious high-density materials and a very high thermal conductivity [8], leading to the need to establish a comparison between conventional and lightweight materials to determine their effects on energy savings in buildings. There are different dynamic simulation techniques to obtain quantitative information to assess the impact and propose improvements in the construction envelopes. An essential part of this work is that it will enable us to determine with high precision the thermophysical properties of the materials through reliable measuring techniques and specialized equipment and, as a result, obtain the thermal performance of the building.

Studies have been carried out that show the benefits of using lighter materials, which reduce dead loads in buildings, handling, and packaging costs. Another advantage is the resultant thermophysical properties that these materials present, which are advantageous because they reduce heat fluxes between the envelope and the exterior environment [9–11]. These materials are lightened with additives, mineral and synthetic substances, and some of natural origin [12,13]. There are materials on the market made from lightweight concrete created using additives (without gravel nor reinforcing steel). Those materials offer a change in their properties, basically as a function of the density and low thermal conductivity, such as cellular concrete. This type of concrete has properties that place them as materials with insulating properties and mechanical resistance to support loads.

These materials are manufactured through processes that, in addition to modifying their density and thermal conductivity with the inclusion of a foaming additive, are cured in an autoclave process (AAC: autoclaved aerated concrete) that consists of subjecting the material to high temperatures and pressure to achieve adequate mechanical resistance. These blocks are made of sand, gypsum, cement, lime, aluminum powder, and water to achieve a chemical reaction that generates hydrogen gas, achieved from the calcium hydroxide in the presence of water and aluminum powder. These materials can reach a nominal density of 500 kg/m^3 and an average compression strength of 40.8 kg/cm^2 with thermal conductivities of up to 0.09 W/m·K [14].

Lastly, commercial cellular concrete has a wide range of thermal conductivity values depending on the density. The thermal conductivity of lightweight concrete varies from 0.09 to 0.22 W/m·K and compared against other commonly used materials, and it has reductions of up to 90%.

1.1. Lightweight Construction Materials

Construction materials are generally dense elements; among the most common are concrete, mortar, and all kinds of water-based mixtures, cementitious and additives. The densities for the heavier materials have values above 2000 kg/m^3, depending on the composition. These materials can become lighter if the composition changes, modifying the supplies to prepare the mixtures, either by changing their proportions or replacing some of the heavier elements for lighter ones or by including additives in the mixtures when it is still fresh.

A recent review of lightweight concrete explored and categorized investigations that contribute to the state-of-the-art in this field [15]. The incorporation of different types of waste materials into foamed concrete was analyzed. It was found that incorporating a maximum of 75% of waste material can be used as an additive in concretes, noticing that a high replacement can interfere with the hydration process. However, the thermal conductivity can be lowered to the range of 0.1–0.7 W/m·K, which is low compared to conventional construction materials. Finally, while incorporating waste materials in foamed concrete will decrease the compression strength, fairly attractive materials can be achieved.

The structural performance of lightweight EPS-foam (LEPSF) concrete was studied for its application in slab systems [16]. The analyzed LEPSF concrete had a 35 MPa at a 1980 kg/m^3 density, and it revealed excellent structural behavior. In another investigation [17], recycled concrete was developed to mitigate aggregate depletion and help with concrete waste disposal. Industrial waste was used to obtain a material of 1733 kg/m^3 density and good mechanical properties. A new cellular concrete with densities of 500, 700, 800, and 1000 kg/m^3 was created using a synthetic polymer foaming agent [18]. The new mixture properties' evaluation showed a compression resistance of over 60kg/cm^2.

Wagh et al. experimentally investigated cellular concrete of three different densities (700, 1000, and 1400 kg/m^3) using various additives [19]. The conclusions point out that an attractive thermal conductivity value was achieved; nonetheless, the mechanical properties were not studied. Font et al. investigated ecological alternatives to cellular concrete technology to contribute to a necessary shift in the circular economy [20]. One-part alkali materials (AAM-OP) and new alkali-activated materials (AACC) were combined with cement, aluminum powder, and residues to propose novel mixtures. Densities of 660 kg/m^3 showed a compression resistance of 64 kg/cm^2 and a thermal conductivity value of 0.20 W/m·K. The experimental work was complemented with a life cycle analysis finding reductions of 96% on the kgCO_2eq per cubic meter of material compared against traditional materials.

Sodangi and Kazmi carried out a comprehensive analysis of coconut palm wood (CPW) for sustainable construction processes [21]. In the study, 13 impediments were listed and evaluated, which are pointed out as a key factor in the slow adoption of the material. Among the main obstacles were the high processing costs, low market demand, and reluctance to use the CPW due to the perceived low social status. Negro et al. explored various interventions for the application of novel sustainable materials in energy and seismic retrofits [22]. An easy to apply biocomposite was analyzed within the study, demonstrating an improvement in seismic performance and reducing the energy requirements of historical Italian buildings. In the investigation, laboratory tests and noninvasive methods such as thermography were used to understand the capabilities to improve the energy performance concerning the seismic retrofits without compromising the integrity of the building.

One of the main characteristics of the lightened materials is the decrease in their density. Thus their thermal conductivity [23] makes the lightened materials, as they have less weight, are workable, and demand less stress to the structures, generating benefits in logistics and costs in the construction sites. In addition to reducing weight, a low thermal conductivity is critical for the material's thermal performance. Additionally, a reduction in the heat flux through the building envelope will reach adequate thermal comfort and less energy demand due to HVAC.

This work will refer to cellular concrete as a lightened material which consists of a mixture of cement, water, and sand with a significant volume of voids through the incorporation of air with the use of foam, that generally is used to manufacture panels and to divide walls with or without mechanical load. The recommended foam density to achieve good stability is 60 kg/m^3 [24]. It is generated using an additive and an instrument that converts the additive into foam using pressurized air. Its proportion is vital to consider the absolute volume that the foam will occupy within the mixture and adjust through laboratory tests.

Cellular concrete can be an environmentally friendly material with excellent insulation and low-density properties (300–1800 kg/m^3) that yields moderate mechanical performance [25]. Cellular concretes are increasingly used in applications in the construction industry [26]. According to Chica and Alzate [27], there are two methods of manufacturing the cellular concrete depending on the origin of the material that generates the mixture's voids. The first is based on the chemical reaction, and the second is based on the manufacturing of foam, which is a generalized and commonly used method. However, it requires the appropriate equipment or machinery to ensure a dry foam, which is well integrated into the concrete mixtures.

Measurements of mortar and concrete mixtures' mechanical properties, both fresh and hardened, are common in construction materials laboratories. However, measures related to thermal performance are different, such as the case of thermal conductivity [28], since it requires specialized equipment if measurements are to be carried out with a low error. This work presents the resulting thermophysical values of the designed materials with high certainty levels, which allows estimating the building's thermal performance through dynamic simulations with a closer approach to reality.

1.2. Impact of Lightened Materials on the Thermal Performance of Buildings

The thermal performance of buildings, whether residential or nonresidential, depends on several factors; among others is the conformation of the thermal envelope that consists of walls, roofs, floors, doors, and windows. The crystal envelope consists of glazing elements such as windows, domes, and skylights, whereas the opaque envelope is the elements that include walls, floors, and roofs. When talking about lightened materials, this study refers to opaque materials for the construction of walls. Walls can be composed of several materials and layers. To calculate the global thermal resistance, it is crucial to know all materials' thermal conductivity, thickness, and geometric configuration. When buildings are located in extreme climates, whether due to cold or heat, these building systems' thermal responses are significant for achieving thermal comfort and reducing energy consumption.

There are studies of thermophysical properties of various materials that provide information for the design of opaque elements in buildings; one of the investigations worked with materials from agricultural residues to replace cement and fine aggregates [29]. Eiras et al. [30] studied the incorporation of granulated rubber to prepare lightened mortars, resulting in lightening the mixtures and good mechanical properties.

In Mexico, conventional construction systems are based on hard materials such as simple concrete, reinforced concrete, sand-based masonry, and mixtures with binders such as cement and plaster combined with additives. The insulation of construction systems is carried out mainly by adding industrialized insulation materials whose technical specifications are determined in the existing market. However, the lightening of the aforementioned massive systems is not common due to the lack of technology or knowledge in the construction sector. This work proposes designing a series of lightened mixtures using foam and fabricating nonindustrialized cellular concrete on-site, without going through the autoclaving procedure.

The Mexican codes related to the production and testing of concrete mixtures used in this investigation refer to lightweight concrete in general, with a particular subsection for cellular concrete. Complimentary international codes as ASTM C1693-11 [31] for autoclaved aerated concrete (AAC) and ACI-523.3R-14 [32] for cellular concrete were also considered. ASTM-C-1693-11 refers to the characteristics of cementitious products with the inclusion of a foaming agent in steel molds. Whereas in ACI-523.3R-14, critical aspects of cement-based materials proportions, curing, and physical properties of the mixtures are described.

In México, the standard NOM-018-ENER-2011 has the objective to establish the testing methodology for thermal resistance evaluation of national materials with insulating properties [33]. NMX-C-460-ONNCCE-2009 regulates the thermal resistance specifications that apply to residential buildings [34]. Its purpose is to improve the housing habitability and reduce the energy demand related to heating and cooling systems considering the corresponding thermal zone; this code is in concordance with ISO-10456 [35]. All of the mentioned codes are related to the design and production of lightweight materials characterized to have low density and low thermal conductivity. Correspondingly, the regulations help to classify the materials as insulating.

On the other hand, two codes that focus on the envelope and are intended to limit the heat gains in buildings are NOM-020-ENER-2011 [36] and NOM-008-ENER-2001 [37]. Their primary purpose is to reduce the energy demand for cooling; the first code has applications in residential buildings, whereas the second in nonresidential buildings.

1.3. Hot Climate Energy Requirements and Codes

In 2008, air conditioning equipment represented 19.7% of the electricity consumed in the residential sector [38]. Regarding 2018, 30% of the residential sector's energy consumption was used to achieve adequate comfort levels in hot climates [39]. To reduce energy consumption in the residential sector, various public policies for efficient use and energy savings have been implemented in Mexico. These efficiency measures have allowed an average user's consumption in temperate climates to decrease since 2001 until reaching consumption levels similar to those of 1989 in 2014. On the other hand, the consumption of average users in hot climates has increased by 19.4% for this same period (1989–2014). These differentiated phenomena are explained, to a large extent, because the most effective public policies have been aimed at improving the efficiency of electrical equipment (lighting and refrigerators). In contrast, they have made very little progress regarding the envelope of buildings, which is what determines in a more significant way the energy consumption in the houses located in regions with a hot climate [40].

The NOM-020-ENER-2011 criterion is based on comparing the heat gain of a projected building, which must be less than a reference building. The heat gain of the reference building is established in this standard and depends on the climatic zone where the construction is planned. This code does not specify a minimum R-value for the materials that comprise the walls or ceilings, but this physical property must be considered to calculate heat gains. On the other hand, the standard NMX-C-460-ONNCCE suggests R-values for roof and walls in three categories (minimum habitability and energy savings) for each climate zone [34]. However, this standard is not mandatory in comparison to NOM-020. Although NOM-020 does not establish an R-value, this property can be determined indirectly from the heat fluxes considered for the design of the reference building. In the particular case of Hermosillo, this reference value is 2.1 m^2K/W. In comparison, NMX-C-460-ONNCCE suggests a minimum R-value of 1.00 m^2K/W and qualifies as an energy-saving level when it presents an equivalent value of 1.40 m^2K/W or higher.

Although the current regulations influence the minimum R-value of the walls and ceilings of new homes, they have not been established in the present reality due to various factors. However, some rules and programs allow obtaining social housing subsidies and more generous financing at more affordable interest rates. They include envelope elements as part of a series of "eco-technologies" that must be considered to grant these kinds of support. Thus, these programs are the real factor that promotes insulating materials in social housing construction [40].

1.4. Energy Use Increase Due to Climate Change

The conventional approach of using Typical Meteorological Year (TMY) data to analyze and quantify buildings' energy consumption has been used and discussed for several years. Various investigations indicate that buildings' thermal and energy performances in the long term are not adequately represented with the commonly used TMY. Researchers emphasize the importance of creating future weather data that considers the yearly variations due to the nature of the standard TMY, neglecting the annual variations caused by the weather changes. In the investigation by Zhu et al. [41], they modified the TMYs using the Morphing method to calculate the energy demand of buildings in Shanghai, taking into consideration the RCP4.5 from the Intergovernmental Panel on Climate Change (IPCC). Huld et al. presented a methodology to generate TMYs based on satellite data; then, the weather data was validated against 487 European meteorological stations [42]. The authors mention that the TMY concept, conceived more than 30 years ago, could be replaced by using time series with 10 years or more data and that this will result in accurate information and faster results. Their standard deviation indicates that a better approach could be obtained by applying a more extended time series rather than employing regular TMY data.

In the research by Hosseini et al., a machine learning method is presented to improve the weather files required for the energy performance analysis on buildings, taking into account climate change scenarios [43]. The investigation proposes a methodology for general circulation models (GCM) to estimate the hourly future building energy performance. The quantile-quantile method was applied to diminish the data bias to fit the GCMs to a specific geographic location. Afterward, a hybrid classification-regression model was applied to reduce the corrected GCM data. The designed workflow utilizes observed weather data to find similar historical weather data patterns and then use it to generate future weather data sets.

Many studies have performed energy and thermal evaluation of buildings to confirm the future use of modified TMYs. Farah et al. calculated the future energy consumption of buildings to increase the cooling requirements between 29% and 31% and a decrease in the heating requirement between 21% and 22% [44]. Pyrgou et al. [45] performed a study to contrast measured weather data from selected weather stations in Perugia to understand the urban heat island (UHI). As a result of this investigation, the authors state that their future work will be centered on creating weather files that consider the impact of the appearance of heatwaves and cold waves, UHI, and other climate change effects. Chakraborty et al. [46] emphasize that accurate weather information is vital to estimate buildings' energy performances, which is currently achieved using TMY files as the TMY3 file, conceived by the Department of Energy (DOE). The TMY3 represents past climate behavior, and it has been found that it does not adequately estimate the economic feasibility and future energy performance. This investigation introduces a novel hybrid modeling methodology that estimates the probability and establishes a machine regression to predict long term performance. Tianzhen Hong et al. [47] performed a large-scale simulation of three office buildings to quantify the impact on the peak electricity demand for 30-year weather data. This study's particular conclusions are that the annual climatic variation affects the peak electricity demand compared to the overall building's energy consumption.

Another issue studied is the impact of the electric grid's energy demand over the years [48]. The electricity demand differences were modest at the early projection time; nevertheless, the changes become more substantial in the last 10 years of the analysis. The cooling demand leads to an increase in energy consumption and the temperature in winter and summer. Clarke et al. [49] found that global energy demand increases by 0.1% for every 2 °C increase in temperature. The results also indicate that the net demand differences are not homogeneous worldwide; for regions where heating requirements are higher, the expectation is that cooling demands will increase. An analysis made within the European scope shows that temperature increase is imminent [50]. As a consequence of climate change, a 2090 scenario shows growth between 50.8% and 119.7% of annual heating and cooling demand. Wang and Chen [51] also found a relationship between energy use and climate change effects depending on the geographic zone. In the U.S., by 2080, an overall energy decrease for climate zones 6 and 7 and an increase in the climate zones 1 to 4 will occur. It was found that by 2080, passive cooling will have a positive impact in San Francisco and Seattle, but it will not be satisfactory in San Diego.

Resch et al. 2021 proposed a dynamic life cycle assessment (LCA) methodology that addresses temporal aspects and future technological improvements [52]. Additionally, a case study was included where 20 buildings were analyzed through the presented technique. The buildings' operations were analyzed considering the climate change impacts related to materials from the production stage, transport, and end-of-life aspects. The estimated metrics expose the implications and drivers required to generate the statistical emission profiles. The encountered results found that the building lifetime, the time horizon, and the construction waste are the most sensitive parameters to mitigate the associated carbon emissions in construction materials.

1.5. Assessment and Early Design Strategies in Residential Buildings to Mitigate Climate Change

Early design approaches and interventions have been mentioned as key to successfully contribute to climate change issues in residential buildings and detached homes. Karimpour et al. [53] evaluated the changes in heating and cooling due to climate change in Adelaide, Australia. Although there is currently a predominant heating demand in the studied city, according to the analysis, by 2070, the city will predominantly require energy for cooling. In the study, the decisive strategies in housing are high insulation levels, increased roof insulation, and highly reflective roof coatings. On the other hand, glazing appears to be one of the most influential parameters due to all of the evaluated houses obtaining ratings of 7 stars or more (1–10, the fewer stars it means a better energy performance). Figueiredo et al. [54] performed an analysis of the energy demand modification in Portugal's residential buildings. It was found that the energy requirements will increase from 5% to 60%, and the appliances were identified as the main drivers for the energy demand. Likewise, in Hong Kong [55], an increase in the cooling requirements among 12.3% and 21.6% of 2017–2100 was found. An increase in the interior temperature was a costless and straightforward strategy as an adaptive thermal comfort approach.

The energy consumption of residential buildings in a Mediterranean climate was calculated using TRNSYS for 2048–2052 and 2096–2100 [56]. The findings showed that the most effective actions are reducing infiltrations and increasing levels of insulation. Invidiata and Ghisi [57] obtained the change in Brazil's residential buildings' energy requirements using the IPCC A2 scenario. It was found that the cooling needs will increase around 56–112% (2050) and 112–185% (2080), while a decrease of up to 94% on the heating requirements by 2080 can be expected. The use of the proposed passive strategies could reduce the energy requirements by up to 50%.

A study in Chile showed that according to climate change, the mean temperature is expected to rise between 0.68 and 1.51 °C by 2050–2065 [58]. The results point out that the heating demand will decrease between 13% and 27%; likewise, changing the current regulations must be considered due to the foreseen future failure in compliance. Flores-Larsen et al. [59] analyzed the heating and cooling energy consumption in Argentinian representative housing applying the climate change A2 scenario. The authors found that the heating requirement will increase around 6.0–7.6 kWh/m^2, while the cooling will increase by 1.7–8.4 kWh/m^2. Finally, Gerek and Arsan [60] explored the increase in energy consumption and CO_2 emissions related to heating and cooling equipment in Turkey's residential buildings by 2080. The analysis showed that the decisions that directly affect the buildings' energy and environmental performances are the ones made on the SHGC and the U-values of the non-opaque surfaces.

This work develops a methodology based on the design of experimental mixtures with cellular concrete, its elaboration, and determination of thermophysical properties, such as fresh state properties, to determine its utility in the constructive activities and dry state to determine the compression strength, thermal conductivity, and specific heat. Subsequently, the evaluation of potential energy savings throughout a year using dynamic simulation and thus quantifying the CO_2 emissions to the atmosphere and a scenario projection of climate change in the future. This research's main contribution is to determine the useful mixture of low density and thermal conductivity and the thermophysical properties experimentally, using specialized equipment. Finally, finding the electric energy savings in the buildings and reducing CO_2 emissions for future scenarios.

2. Materials and Methods

This project was carried out in two stages: experimental and dynamic simulation. The experimental stage consists of the design and manufacture of lightened mixtures according to the corresponding normativity and the measurement of thermophysical properties. Simultaneously, the simulation stage contemplates determining the cases to analyze the prototypical house's related geometry, typical construction materials for the wall, and envelope to compare the results against the proposed materials.

This work begins with designing a mixture of natural mortar (sand–cement–water) in a proportion of 1:4 as a control mixture. A new mix was designed with an additive or foaming agent to reduce the density of the mixture. Once the theoretical design was done, the material was fabricated in the laboratory to test the thermophysical properties such as fluency and air content in a fresh state. The properties related to its rigid state, such as the compression strength, thermal conductivity, and specific heat, were measured for both the control mixture and the lightened mixture.

The second stage contemplates using prototypical social housing [8], developed with traditional constructive systems in the northwest of Mexico (lightened slab, walls made with hollow concrete blocks, and conventional doors, windows, and floors). A series of dynamic simulations were carried out throughout a year to evaluate the building's thermal performance. Subsequently, further simulations were carried out, integrating the designed cellular concrete on the walls. Finally, the electric energy consumed and the CO_2 emissions were compared to determine the savings to use lightened materials against the conventional materials.

2.1. Experimental Method

The project's experimental stage consists of designing a natural 1:4 sand–cement mortar mixture as a control mixture and then lightening it with foaming agents to obtain densities lower than the original. The mixtures were manufactured, and their physical properties were measured in a fresh state. The test samples were then manufactured, and after a 28-day setting (dry state), the lightened samples were ready to determine the compression strength, thermal conductivity, and specific heat using specialized equipment for six different densities (800, 1000, 1200, 1400, 1600, and 1800 kg/m^3); all of the above following current regulations.

From the results, the best samples were selected based on the load capacity and density to be used as construction elements. A natural setting was considered in the manufacturing process, hydrating the mixture with water, a conventional procedure at the worksite, and different from industrial-scale that uses an autoclaving process to produce cellular concrete.

2.1.1. Theoretical Design of a Cellular Concrete Mixture.

Table 1 shows the theoretical proportions of the control and lightened samples. CPC-30-R cement with a density of 3000 kg/m^3 graded natural sand and contaminant-free water. The sand was studied using granulometric analysis to determine its properties according to the standard NMX-C-077-ONNCCE-1997 [61]. These analyses showed that the sand has a density of 2560 kg/m^3 and an absorption percentage of 1.5%, a contamination percentage of 5.8%, a fineness modulus of 2.8, and 4.5% loss on washing [62]. The recommended foam density to achieve good mix stability is approximately 60 kg/m^3. This is assured by using an additive and an instrument that converts the additive to the foam using pressurized air. The volume that the foam will occupy within the mixture must be considered and adjusted through laboratory tests for its proportioning.

Table 1. Proportion of cellular concrete for different densities in a volume of 15 L and percentage of fluidity.

Property	Control	C1	C2	C3	C4	C5	C6
Mixture density (kg/m^3)	2135	1800	1600	1400	1200	1000	800
Mixture volume (m^3)	1.000	0.8432	0.7496	0.6569	0.5622	0.4685	0.3748
Cement CPC 30R (kg)	6.225	5.243	4.661	4.078	3.495	2.913	2.330
Sand (kg)	21.225	17.896	15.91	13.91	11.93	9.942	7.95
Water (kg)	4.575	3.861	3.43	3.00	2.574	2.145	1.71
Foam (kg)	0	0.141	0.22	0.31	0.39	0.478	0.563
Fluidity (%)	106	124	125	125	122	122	122

The cellular concrete volume ratio concerning the original mortar's total volume was determined to obtain a cellular concrete mix of specific density. This was achieved by dividing the desired density by the density of the control mix; in other words, the foam will fill the difference between these volumes; according to these criteria, the proportions were calculated to obtain cellular concrete of different densities, with a fluidity range of 105 ± 15%.

Conventional laboratory equipment and tools such as a mixer, molds for samples, basic tools, and test equipment to obtain the mortar characteristics in a fresh state were used for the manufacture and tests of the samples. Initially, the control mixture was fabricated with natural cement–sand mortar in a 1:4 ratio, according to the dosages presented in [62].

Cellular concrete lightened mortars were also manufactured, adding the foaming agent according to the proportions shown in Table 1, which correspond to the densities of 800, 1000, 1200, 1400, 1600, and 1800 kg/m^3. The brand of the additive used to generate the foam is BUINY A.E./C.C.100, a foaming additive.

The equipment used to generate preformed foam for the cellular concrete is called: UTC 400-E/simple. Once the mixtures were made, their properties in the fresh state were measured in the laboratory; in relation to the fluidity and according to NMX-C-144-ONNCCE-2015 [63], the tests determined that the mixture presents an adequate fluidity and an air content of 15–70%. Then, three cubic samples of 0.05 × 0.05 × 0.05 m were manufactured both for the control sample and for each of the cellular concrete samples of different densities; they were tested to determine compression strength after 28 days, according to the standard NMX-C-061-ONNCCE-2015 [64]. Additionally, samples with dimensions of 0.15 × 0.15 × 0.04 m were manufactured to measure thermal conductivity and specific heat.

2.1.2. Manufacture of Samples and Construction Components

According to the current regulations, after manufacturing the mixture, 0.05 × 0.05 × 0.05 m samples were produced to carry out the compression tests (Figure 1a). Additionally, 0.15 × 0.15 × 0.04 m samples were made for the thermal conductivity, specific heat, and density tests (Figure 1b). Once the properties were determined, 0.12 × 0.2 × 0.4 m blocks were manufactured (Figure 1c).

(**a**) Compression test probes: 0.05 × 0.05 × 0.05 m

(**b**) Probes for thermal conductivity, specific heat and density tests: 0.15 × 0.15 × 0.04 m

(**c**) Produced cellular concrete blocks: 0.12 × 0.2 × 0.4 m

Figure 1. Test samples and produced cellular concrete blocks.

The proposed wall constructive systems are based on nonstructural cellular concrete blocks of 0.12 × 0.20 × 0.40 m with a 1600 and 1800 kg/m^3 density. The proposed materials are within the range, according to Mexican regulations [65], lightened mortars of a density of 1600 to 1800 kg/m^3 can be used on masonry with a compression strength among 60 and 100 kg/cm^2, respectively. Likewise, according to the NMX-C-486-ONNCCE-2014 [66], a type II mortar with a compression strength value equal or higher than 125 kg/cm^2 is recommended to attach the blocks.

Compression resistance of 40–45 kg/cm^2 was determined through interpolation for the 1600 and 1800 kg/m^3 designed lightweight blocks, respectively. An additional 800 kg/m^3 mortar is proposed for improving lightness and insulation; the use of this material is subject to its use as a filler in nonbearing sections.

2.1.3. Measurement of Thermophysical Properties in the Dry State

After the 0.05 × 0.05 × 0.05 m cubic samples were set, the compression strength was measured at days 3, 7, and 28 by triplicate for each of the studied densities. The mean value of the measurements was used for the calculations. A semiautomatic Digimax Dual Chamber with 33/15 kN capacity of compression and flexion was used for the compression strength measurement on the cellular concrete samples following ASTM C109/C109M [67].

The thermal conductivity and density of the 0.15 × 0.15 × 0.04 m were measured with the guarded hot plate method. The probes were subject to three temperature gradients, and then a polynomial regression was performed to obtain the thermal conductivity value. A Guarded Hot Plate Apparatus Lambda Meter Ep 500e was used for the measurement of thermal conductivity (W/m·K) with an accuracy between 0.7% and 1% and reproducibility between 0.2% and 0.5% in accordance with ISO 8302:1991 [68], ASTM C177-19 [69], EN 1946-2 [70], UNE-EN 12664:2002 [71], UNE-EN 12667:2002 [72], UNE-EN 12939:2001 [73], and DIN52612 [74].

To measure specific heat (MJ/m^3·K), probes of 0.15 × 0.15 × 0.04 m were drilled for the sensor placement. For this purpose, a KD2-PRO Thermal Properties Analyzer, Decagon Devices Inc., was used with a sensor SH-1 for hard materials with an accuracy of ±10% by the standard ASTM D5334-08 [75]. In Figure 2, the utilized equipment for compression strength (a), thermal conductivity (b), and specific heat (c) can be observed.

(**a**) Compression strength: Digimax Chamber.

(**b**) Thermal conductivity and density: EP-500.

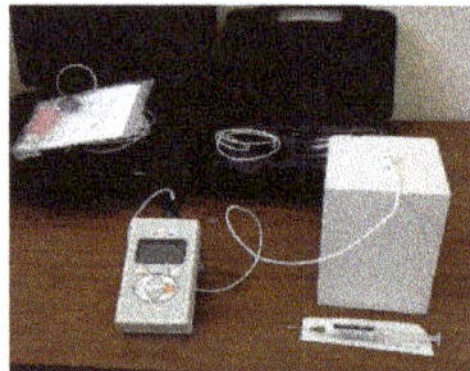

(**c**) Specific heat: KD-2-PRP.

Figure 2. Equipment used in compression strength, thermal conductivity, and specific heat measurements.

2.2. *Dynamic Simulation*

The software TRNSYS 17 was used to determine the energy performance of the building. This building is located in the city of Hermosillo, Sonora, with a BWh climate which corresponds to a hot desert climate, according to the Köppen–Geiger classification. The software Meteonorm provided the weather file for a typical year, and commonly used materials were considered for the house envelope constructive systems.

For this study, the thermal comfort range was calculated according to the ASHRAE 55-2004 standard [76]. The calculated temperature range was 20.6–27.3 °C, and it was used to operate the heating and cooling systems with the type 56 from TRNSYS.

The simulation uses a simplified heating and cooling energy requirement estimation through the type 56 mentioned above. The premise is that the cooling and heating demand directly depends on a defined indoor temperature range. Therefore, the required output power to maintain the indoor temperature within the comfort range results in a positive value for the cooling energy and a negative value for the heating necessities. This is related to the convention that heat needs to be added or removed from the studied building.

2.2.1. Physical Model of the House and Location

The studied house represents a typical construction of social housing in the northwest region of Mexico [8]. The house is attached to the neighboring houses, while the main facade is exposed to the outside. The house's facade is oriented to the south and has a floor to ceiling height of 2.5 m; it has a backyard, and an interior small service patio, one of the rear walls is also adjacent to the posterior home. The house was built on a land of 90 m^2 and has a construction area of 55.8 m^2, and has a floor area/window wall ratio of 0.07. The home is divided into the following thermal zones: (1) North Bedroom, (2) South Bedroom, (3) Bathroom, and (4) Kitchen-Living Room. The house layout and facade from the studied case study can be observed in Figure 3.

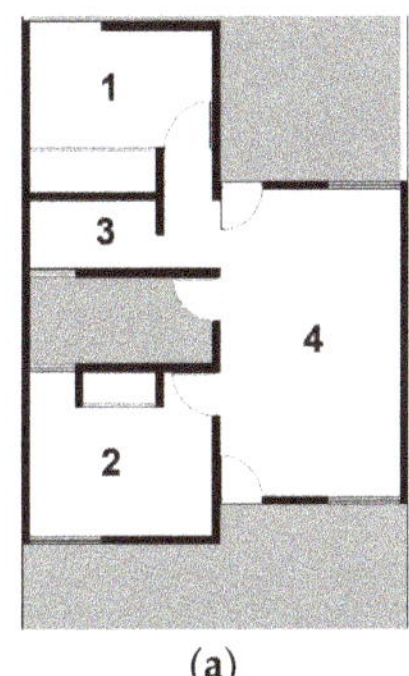

(**a**)

(**b**)

Figure 3. Analyzed case study. (**a**) Floor plan layout, (**b**) facade.

The investigated case study was chosen due to two crucial parameters: extremity on the selected climate and the housing typology classification. The studied city is located in extreme weather; the situation makes it critical to rethink this housing design and propose affordable materials with appropriate thermal properties. In 2015, the population census counted that the state of Sonora has 812,567 households, from which 257,537 are located in the city of Hermosillo, and 99.36% of them had electricity access [77]. On the other hand, Hermosillo consumes 32% of the electricity of the state, and residential buildings located in the city use 13.5% of the annual consumed electricity [78].

The selected house corresponds to a dwelling classified as "popular housing," according to the Mexican National Housing Commission (CONAVI) [8]. Such classification considers six types of homes; the smaller three are considered as "social housing" and corresponds to the economic (40 m^2), popular (50 m^2), and traditional (71 m^2) housing. While the remaining three are the medium (102 m^2), residential (156 m^2), and residential plus (+188 m^2) houses.

The analyzed building belongs to a housing segment of low-income families and, therefore, the least expensive option on the market. The distinctive low-cost social housing built in Sonora and Mexico has helped massive construction, thus benefiting millions of families. Nevertheless, they are made with basic materials and have a low-quality thermal performance in many climatic circumstances. With the demographic increase and the subsequent surge in housing demand, the immediacy to improve the housing quality within the context of climate change, especially for the most vulnerable sector, is seen as a more than appropriate action.

2.2.2. Climatic Conditions of the Arid Weather

The region of study is considered part of the dry weather (B), according to the Köppen–Geiger classification. This weather presents scarce precipitation, which incites a broad thermal gradient between day and night. It is possible to reach a maximum temperature of up to 50 °C while it drops below −10 °C at night.

The IPCC has six scenarios, with one group in the A2, B1, and B2 families and three within the A1. In the A1 family, the following scenarios were set. The A1F scenario consists of the intensive use of fossil fuels, A1B, a balanced scenario, and the A1T with a predominance of non-fossil fuels. On the other hand, the A2 scenario represents a scheme where regional economic development is more substantial. The technological change is slower; the coal is again employed, and oil and gas shares are reduced. In this work, additionally to the typical TMY, information from 2030 and 2050 is also integrated, considering a balanced scenario from the A1 storyline (A1B) and the A2 scenario from IPCC [79,80]. Figure 4 shows monthly averages of temperature (T_{amb}, °C) and the global horizontal radiation (Gh, W/m^2) of the historic TMY (T_{amb}: 2000–2009, and Gh: 1991–2010), and the A1B and A2 scenarios.

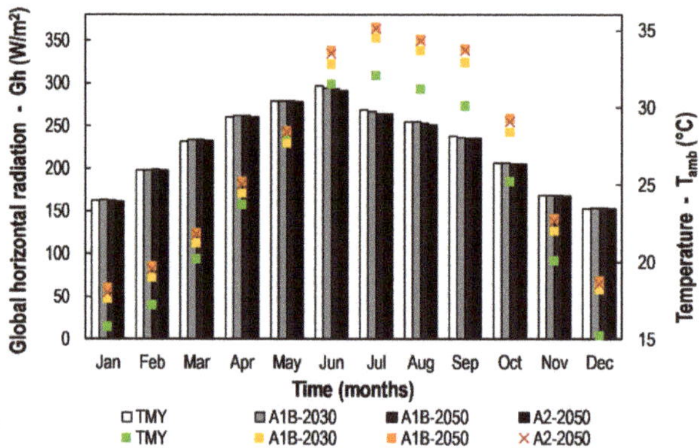

Figure 4. Temperatures and global horizontal radiation.

Table 2 shows a summary of the performance in the radiation, the average and maximum temperatures for the periods considered. According to scenario A1B, an increment of average temperatures (T_{avg}) of 1.84 °C by 2030 and 2.61 °C by 2050. In comparison, reaching maximum temperatures (T_{max}) of up to 46.1 °C in 2030 and up to 45.4 °C by 2050, giving, as a result, an increase in the average temperatures from 2000–2009 of up to 3.7 °C (2030) and 3.0 °C (2050). Within the A2 scenario for the year 2050, the minimum temperatures will increase up to 2.2 °C compared to the TMY from 2000–2009.

Table 2. Summary of the global horizontal radiation and temperatures.

	Global Horizontal Radiation (Gh, W/m^2)				Temperatures (°C)			
	TMY	**A1B-2030**	**A1B-2050**	**A2-2050**	**TMY**	**A1B-2030**	**A1B-2050**	**A2-2050**
Average	226.55	226.37	225.62	224.74	24.15	25.99	26.75	26.59
Maximum	1206.00	1111.00	1156.00	1124.00	42.40	46.10	45.40	46.50
Minimum	/	/	/	/	3.00	4.50	5.80	5.20

2.3. Case Studies

According to the test results of the lightened mortar, the composition of three constructive systems was proposed. To evaluate the energy savings, the proposed wall systems were compared against a conventional concrete block wall. The house walls are composed of three layers: cement plaster, block, and gypsum plaster, where the block is replaced according to the analyzed systems.

The incorporation of cellular concretes in the manufacturing of construction elements is based on density. According to Ni Frank Mi-Way et al. [81] and Sari and Sani [82], the densities of 1600–1800 are recommended to manufacture slabs and other load elements where greater strength is required, such as partitions for home construction. On the other hand, materials with densities of 800 or less are proposed for fillings.

Table 3 shows the considered blocks for each study case, representing different constructive systems for the wall. Case 0 is the base case where the walls are constructed with a typical hollow concrete block of 0.12 m of thickness with conventional plaster, lightened slab, with traditional doors and windows. Cases 1 and 2 will vary the composition of the walls, for which the substitution of blocks is made with cellular concrete. Case 3 represents a highly lightweight filler cellular material. In all cases, the house is constructed with a lightened slab composed of white waterproofing, concrete, lightened slab, and gypsum plaster, with an R-value of 0.8 [62].

Table 3. Selected systems for dynamic simulations.

Case	Wall Composition	Mixture (Agreement with Table 1)
Case 0	Concrete hollow block	Control
Case 1	Solid cellular block (1800 kg/m^3)	C1
Case 2	Solid cellular block (1600 kg/m^3)	C2
Case 3	Filler cellular material (800 kg/m^3)	C6

3. Results

3.1. Thermophysical Properties

Table 4 shows the results of compression strength, thermal conductivity, and specific heat tests for the hardened state of the cellular concrete mixtures.

Table 4. Result of thermal conductivity measurements, compression strength, and specific heat for the dry mixtures.

Property	C1-1800	C2-1600	C3-1400	C4-1200	C5-1000	C6-800
Density (kg/m^3)	1810	1610	1457	1216	1065	863
Thermal conductivity (W/m·K)	0.2958	0.2517	0.2480	0.2057	0.1682	0.1399
Compression strength (kg/cm^2)	88.8	77.5	54.3	18.6	10.4	8.3
Volumetric specific heat (MJ/(m^3·K)	1.56	1.64	1.14	1.34	0.92	1.72

According to the results shown in Table 2 regarding the compression strength values, it is determined that those with densities of 1800 and 1600 comply with a value greater than 75 kg/cm^2, according to NMX-486-ONNCCE -2014 standard [66]. It can be observed that a decrease in thermal conductivity is correlated to a reduction of the density. The effectiveness in the reduction of the thermal conductivity as a material with insulating properties will be evaluated in the next section of this paper, throughout a dynamic simulation of a building.

According to all of the above, a contrasting analysis is proposed using conventional materials as a base case against lightened cellular concretes with densities of 1800 and 1600. Besides, a study is carried out with the cellular mortar with a density of 800, which can be used as a confined filling material.

In Table 5, the R-values of the evaluated systems are shown. Here, it can be observed that the R-value of the analyzed cases increases from 0.2874 (Case 0) to up to 0.9651 for the less dense mixture (Case 3). The data in Table 4 was used to carry out the dynamic simulations to determine the energy use and thermal comfort of the cases listed in Table 5.

Table 5. Analyzed systems.

Case	Wall Composition	R_{value} (m^2·K/W)
Case 0	Concrete hollow block	0.2874
Case 1	Solid cellular block (1800 kg/m^3)	0.5131
Case 2	Solid cellular block (1600 kg/m^3)	0.5841
Case 3	Filler cellular material (800 kg/m^3)	0.9651

3.2. Dynamic Simulations

In this section, the temperatures that occur inside the house are presented. Additionally, the annual energy demand related to heating and cooling systems (kWh) and the CO_2 emissions associated with the systems' use is given.

3.2.1. Indoor Temperatures

Table 6 shows the annual average, maximum, and minimum indoor temperatures that take place inside the house when no HVAC system is operated. The results were obtained from the cases analyzed, representing a traditional concrete block wall (Case 0) and the other three cases (1–3) corresponding to the three different densities of the cellular concrete studied. Likewise, the temperatures that would occur in the scenarios (A1B) of climate change for the years 2030 and 2050 are presented, as well as for the heterogeneous scenario known as A2 for 2050. In Table 6, values in bold black letters were marked to represent average temperatures that are considered as too high and completely out of the thermal comfort range. Additionally, in the maximum and minimum temperatures, the numbers in red and blue bold letters were added to point out very high (in red) and very low (in blue) indoor temperatures.

Table 6. Annual temperature behavior without air conditioning systems.

	Case	TMY	A1B-2030	A1B-2050	A2-2050
Average temperatures (T_{avg}, °C)	Case 0	26.32	**28.04**	**28.80**	**28.61**
	Case 1	26.47	**28.20**	**28.95**	**28.76**
	Case 2	26.51	**28.23**	**28.99**	**28.80**
	Case 3	26.67	**28.39**	**29.14**	**28.95**
Maximum temperatures (T_{max}, °C)	Case 0	39.13	41.55	42.48	41.59
	Case 1	38.85	41.24	42.16	41.22
	Case 2	38.69	41.07	41.98	41.09
	Case 3	38.41	40.75	41.62	40.98
	Case 0	39.13	41.55	42.48	41.59
Maximum temperatures (T_{min}, °C)	Case 0	11.07	12.58	13.54	13.07
	Case 1	11.77	13.25	14.24	13.74
	Case 2	12.01	13.52	14.49	14.00
	Case 3	12.62	14.18	15.10	14.65
	Case 0	11.07	12.58	13.54	13.07

It is observed that when using the TMY with historical information, the average temperatures occurring inside are adequate. However, for all future scenarios, the average annual temperatures that are registered are greater than 28 °C and considered out of the comfort range. On the other hand, the maximum temperatures reached inside the house range between 38.41 °C (Case 3, TMY) and 42.48 °C (Case 0, a1b-2050). It is also noted that cellular concrete's application manages to reduce, on average, 0.75 °C. This maximum temperature occurs inside the house when the temperatures are compared between the case with a conventional block (Case 0) and the lighter cellular concrete (Case 3).

Similarly, when observing the house's minimum temperatures, the lower would be 11.07 °C (Case 0, TMY) and 15.10 °C (Case 3, A1B-2050). The average difference between the conventional block (Case 0) and the cellular concrete of Case 3 would be 1.57 °C.

3.2.2. Thermal Comfort

Table 7 shows the promoted mean vote (PMV) of the houses with no HVAC systems, finding similar tendencies as those found on temperatures. The PMV values were indicated in bold black, red, and blue color. Where the bold black PMV value represents a warm sensation, and the red and blue connote an inappropriately high and low PMV.

Table 7. Averaged promoted mean vote (PMV) without air conditioning systems.

	Case	TMY	A1B-2030	A1B-2050	A2-2050
Average PMV	Case 0	0.39	0.80	0.98	0.94
	Case 1	0.43	0.84	**1.02**	0.98
	Case 2	0.44	0.85	**1.03**	0.99
	Case 3	0.48	0.89	**1.07**	**1.02**
Maximum PMV	Case 0	3.43	3.99	4.21	4.00
	Case 1	3.36	3.92	4.14	3.91
	Case 2	3.32	3.88	4.09	3.88
	Case 3	3.25	3.79	4.00	3.85
	Case 0	3.43	3.99	4.21	4.00
Maximum PMV	Case 0	−3.42	−3.03	−2.78	−2.90
	Case 1	−3.24	−2.85	−2.60	−2.73
	Case 2	−3.18	−2.78	−2.53	−2.66
	Case 3	−3.01	−2.61	−2.38	−2.48
	Case 0	−3.42	−3.03	−2.78	−2.90

It can be observed that average PMV is mostly around the "slightly warm" condition, especially for the results of the cases related to the year 2050, where values greater than 1.0 were calculated. On the other hand, the maximum PMV values were between 3.21 (Case 3, TMY) and 4.21 (Case 0, A1B-2050). These PMV values would be considered entirely unacceptable from the thermal comfort perspective since it exceeds the "hot" condition, generating extreme heat stress. The minimum PMV values registered are between −2.48 and −3.42, which in the same way would cause extreme cold stress in the occupants of the house.

3.2.3. Annual Energy Demand

The annual energy demand generated in the houses composed of walls with different lightening levels was calculated for the analyzed IPCC scenarios. Figure 5 shows the respective heating and cooling demand. It is observed that the cases where the TMY with historical information is used result in lower energy requirements, meaning that out studies carried out on the use of energy in buildings could result in an underestimation of energy requirements necessary to keep a house in thermal comfort. When analyzing the energy requirements in scenarios A1B (2030 and 2050) and A2 (2050), it is observed that, in general, there would be a greater energy demand. The energy demand associated with air conditioning would increase; however, the heating requirement would be reduced over the years.

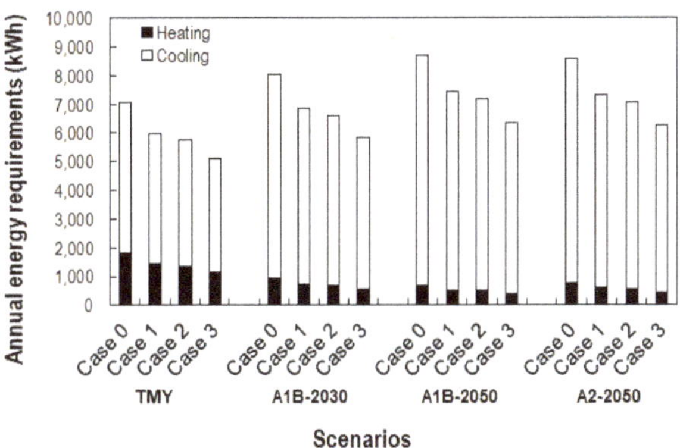

Figure 5. Annual energy requirements per studied case.

This figure clearly shows that applying lightweight cellular concrete would reduce the overall energy requirement of the house. For example, in scenario A1B-2030, when comparing the energy required for the heating and cooling systems in the house for Case 0 (8025 kWh) and Case 3 (5850 kWh), a reduction of 37% is observed. Likewise, the energy demand for cooling the house reduces by 25% for A1B-2030 and A1B-2050. As for the heating, the required energy would be reduced over the years. However, when applying the lightened concrete, it decreases by 43% for both A1B-2030 and A1B-2050 scenarios.

When comparing the energy consumption to cool the house, it was found that these will increase by 35% (A1B-2030), 52% (A1B-2050), and 49% (A2-2050) for the historical TMY and the traditional construction of the house (Case 0). In the same way, energy consumption to heat the house will be reduced, on average, 96% (A1B-2030), 61% (A1B-2050), and 58% (A2-2050).

3.2.4. Reduction of Annual tCO_2e Emissions

The CO_2 emissions calculation was undertaken considering each scenario's annual electricity requirements and based on the Mexican Electricity System "Emissions Factor Report", published in February 2020. The 2019 factor is reported to be 0.505 tCO_2e per MWh [83]. Figure 6 shows the reduction in the annual CO_2 emissions per studied case and scenarios. This figure shows that, as expected, the lower density concrete allows a decrease of CO_2 emissions. For example, for the A2-2050 scenario a reduction of 14.48% of tCO_2e emissions for Case 1 is achieved, 17.52% for Case 2, and 26.88% for Case 3.

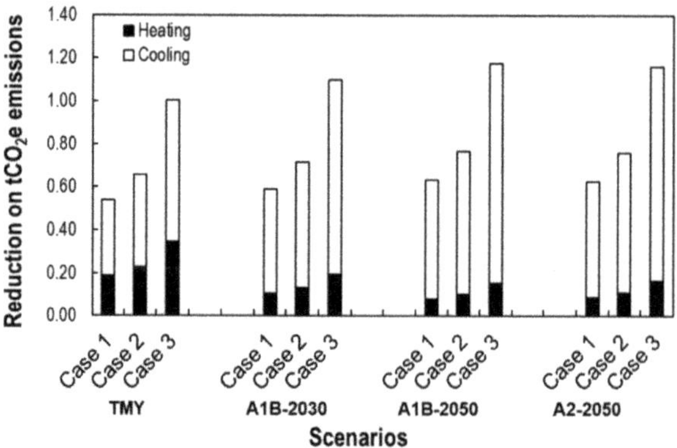

Figure 6. Annual tCO_2e emissions per studied case.

4. Discussion

The residential sector is responsible for the consumption of fair amounts of energy with a substantial environmental impact. The families that live in economic housing in developing countries often face the decision to pay for the energy related to air conditioning systems or live with thermal discomfort. For this reason, in this investigation, lightweight cellular concrete mixtures were proposed for their prospective application on economical housing. The materials were designed and applied to walls to reduce energy consumption due to the use of air conditioning systems in a city with a hot arid climate. The presented methodology's main objective was to acquire accurate information to perform a whole building examination of the material to be recommended for new constructions.

The results showed a reduction in the density and thermal conductivity of the designed materials. The reduced weight and thermal conductivity culminate in energy-saving benefits and the consequent decrease of energy-related CO_2e emissions when applied to a residential building. Thus, this investigation's outcomes show a viable proposal to be immediately used in the construction industry.

The presented methodology can be implemented in a variety of materials and aggregate combinations for numerous building applications. This is especially relevant for on-site materials manufacture, as this practice is scarcely a custom in construction. Additionally, the suitability of implementing measurement with specialized equipment under national and international standards gives certainty to the designed materials' future performance.

The house's thermal performance was calculated through the dynamic simulations, finding very unpleasant indoor conditions when no air conditioning systems are in use. Temperatures of up to 42.48 °C could be achieved inside the home if the traditional concrete blocks continue to be used in construction by 2050. The minimum temperatures are also very low, finding 11 °C in the coldest season of the year. Likewise, even when the average temperatures were analyzed for future scenarios, they are around 28 and 29 °C, which are out of the thermal comfort range.

Similar results were found when the promoted mean vote (PMV) values were explored. For the A1B and A2 scenarios in the year 2050, the averaged PMV is near 1.0, which would result in a "slightly warm" sensation. Even though all of the analyzed mixtures do not result in an adequate average PMV value, the conventional concrete block (Case 0) presents the extreme PMVs. A 4.21 PMV for the 2050 A1B scenario and −3.42 with the historic TMY.

The application of the cellular concrete mixtures results in a lower annual energy requirement. The cooling energy is significantly reduced; nevertheless, a decrease in heating energy was also accomplished. From the yearly energy requirements figure, it can be observed that if the conventional concrete block continues to be used massively for economical housing, the air conditioning energy requirements will increase up to 49% by the year 2030 and 61% by the year 2050 (A1B scenario). Even with the lightweight material application, energy requirements by 2050 will resemble the concrete block energy performance.

The annual energy demand of 7062 kWh was determined for a house located in a hot desert climate through dynamic simulation, where 74% of that energy is destined for cooling. According to the constructed area of the house, this value corresponds to the energy consumption of 126.6 kWh per square meter. This energy consumption is reduced in a range of 15.1–28.1% if the hollow concrete block currently used to construct social housing is replaced with solid blocks of the cellular concrete proposed in this investigation.

If IPCC scenarios are considered, the annual consumption of the base case increases at levels ranging from 143.8 (A1B-2030) to 155.9 kWh/m^2 (A1B-2050). The percentages of energy reduction given by the use of cellular concrete remain constant for these scenarios, but the magnitude of savings increases from a range of 19.1–35.6 kWh/m^2 to ranges of 21.0–31.0 and 22.5–41.7 kWh/m^2 for scenarios A1B-2030 and A1B-2050, respectively.

Finally, through the CO_2e emissions reduction analysis, it was found that the implementation of the novel cellular concrete mixtures accomplishes a fair reduction in the emissions related to the energy use of a relatively small home (55.8 m^2).

The study of the thermophysical properties of construction materials should transit to direct applications. This is key to understand the repercussions that imply their implementation in real environments. The knowledge of the mixture's final thermal behavior has implications not only on the technical level, since the benefits are acute at the economic and environmental level, especially with climate change impacts in sight. The environmental benefits of the performed projections are of great usefulness for current decision-making and significant profits for the future, especially in planning public policy for climate change mitigation.

Author Contributions: Conceptualization, A.C.B.-A. and N.A.R.-M.; Data curation, J.L.-A. and M.N.-T.; Formal analysis, J.L.-A. and N.A.R.-M.; Funding acquisition, A.C.B.-A.; Investigation, A.C.B.-A., M.R.-C., S.C.-B. and N.S.-S.; Methodology, N.A.R.-M.; Project administration, A.C.B.-A. and N.A.R.-M.; Resources, A.C.B.-A. and N.S.-S.; Software, J.L.-A. and M.N.-T.; Visualization, A.C.B.-A. and N.A.R.-M.; Writing—original draft, A.C.B.-A. and N.A.R.-M.; Writing—review and editing, A.C.B.-A., J.L.-A., N.A.R.-M. and M.N.-T. All authors have read and agreed to the published version of the manuscript.

Funding: This research was funded by Comision Nacional de Vivienda (CONAVI), project "CONACYT-CONAVI 2014:S003-236187" and Mexican National Council of Science and Technology (CONACYT), the project "CONACYT-PDCPN-2015:1099". Division de Ingenieria from the Universidad de Sonora funded the article processing charges (APC).

Acknowledgments: Special thanks to Claudia K. Romero-Perez (CIMAV-Durango) for the collaboration on the project. Additionally, we thank the architecture students Juan C. Vela-Carrillo and Ruben Torres-Verduzco (Instituto Tecnologico de Durango). Finally, we would like to thank the Municipal Housing Institute (INMUVI-Durango) for the provided information.

Conflicts of Interest: The authors declare no conflict of interest.

References

1. Ye, B.; Jiang, J.; Liu, J.; Zheng, Y.; Zhou, N. Research on quantitative assessment of climate change risk at an urban scale: Review of recent progress and outlook of future direction. *Renew. Sustain. Energy Rev.* **2021**, *135*, 110415. [CrossRef]
2. Kang, J.N.; Wei, Y.M.; Liu, L.C.; Han, R.; Yu, B.Y.; Wang, J.W. Energy systems for climate change mitigation: A systematic review. *Appl. Energy* **2020**, *263*, 114602. [CrossRef]
3. Cabeza, L.F.; Urge-Vorsatz, D.; McNeil, M.A.; Barreneche, C.; Serrano, S. Investigating greenhouse challenge from growing trends of electricity consumption through home appliances in buildings. *Renew. Sustain. Energy Rev.* **2014**, *36*, 188–193. [CrossRef]
4. Arreortua, L.A.S.; Montaño, A.M.P. Social housing and habitability in the periphery of metropolitan area of the valley of Mexico. *Rev. Geogr. Norte Gd.* **2020**, *76*, 51–69.
5. Gomez, C.G.; Morales, G.B.; Torres, P.R. Sensación térmica percibida en vivienda económica y auto-producida, en periodo cálido, para clima cálido húmedo. *Ambient. Constr.* **2011**, *11*, 99–111. [CrossRef]
6. Martínez-Montejo, S.A.; Sheinbaum-Pardo, C. The impact of energy efficiency standards on residential electricity consumption in Mexico. *Energy Sustain. Dev.* **2016**, *32*, 50–61. [CrossRef]
7. McNeil, M.; Castellanos, S.; De Leon Barido, D.P.; Peréz, P.A.S. *Mexico Space Cooling Electricity Impacts and Mitigation Strategies: Analysis Supporting the Summit on Space Cooling Research Needs and Opportunities in Mexico February 15–16 Casa de la Universidad de California, Mexico City*; USAID: Washington, DC, USA, 2018.
8. CONAVI—Comisión Nacional de Vivienda México. *Código de Edificación de Vivienda*; CONAVI: Ciudad de México, Mexico, 2010.
9. Essid, N.; Eddhahak-Ouni, A.; Neji, J. Experimental and Numerical Thermal Properties Investigation of Cement-Based Materials Modified with PCM for Building Construction Use. *J. Archit. Eng.* **2020**, *26*, 04020018. [CrossRef]
10. Nshimiyimana, P.; Messan, A.; Courard, L. Physico-mechanical and hygro-thermal properties of compressed earth blocks stabilized with industrial and agro by-product binders. *Materials* **2020**, *13*, 3769. [CrossRef]
11. Iștoan, R.; Tămaș-Gavrea, D.R.; Manea, D.L. Experimental investigations on the performances of composite building materials based on industrial crops and volcanic rocks. *Crystals* **2020**, *10*, 102. [CrossRef]
12. Kan, A.; Demirboğa, R. A novel material for lightweight concrete production. *Cem. Concr. Compos.* **2009**, *31*, 489–495. [CrossRef]
13. Chen, Y.X.; Wu, F.; Yu, Q.; Brouwers, H.J.H. Bio-based ultra-lightweight concrete applying miscanthus fibers: Acoustic absorption and thermal insulation. *Cem. Concr. Compos.* **2020**, *114*, 103829. [CrossRef]
14. HEBEL Technical Sheet: Hebel. Available online: https://www.hebel.mx/pdf/fichas-tecnicas/v08.20/BlockSolidoFichaTecnica-v08.20.pdf (accessed on 17 November 2020).
15. Shah, S.N.; Mo, K.H.; Yap, S.P.; Yang, J.; Ling, T.C. Lightweight foamed concrete as a promising avenue for incorporating waste materials: A review. *Resour. Conserv. Recycl.* **2021**, *164*, 105103. [CrossRef]
16. Saheed, S.; Amran, Y.H.M.; El-Zeadani, M.; Aziz, F.N.A.; Fediuk, R.; Alyousef, R.; Alabduljabbar, H. Structural behavior of out-of-plane loaded precast lightweight EPS-foam concrete C-shaped slabs. *J. Build. Eng.* **2021**, *33*, 101597. [CrossRef]
17. Zhou, Y.; Gong, G.; Huang, Y.; Chen, C.; Huang, D.; Chen, Z.; Guo, M. Feasibility of incorporating recycled fine aggregate in high performance green lightweight engineered cementitious composites. *J. Clean. Prod.* **2021**, *280*, 124445. [CrossRef]
18. Kadela, M.; Kukiełka, A.; Małek, M. Characteristics of Lightweight Concrete Based on a Synthetic Polymer Foaming Agent. *Materials* **2020**, *13*, 4979. [CrossRef] [PubMed]

19. Wagh, C.D.; Ranjani, G.; Kamisetty, A. Thermal Properties of Foamed Concrete: A Review. In Proceedings of the 3rd International Conference on Innovative Technologies for Clean and Sustainable Development, Chandigarh, India, 19–21 February 2020.
20. Font, A.; Soriano, L.; Tashima, M.M.; Monzó, J.; Borrachero, M.V.; Payá, J. One-part eco-cellular concrete for the precast industry: Functional features and life cycle assessment. *J. Clean. Prod.* **2020**, *269*, 122203. [CrossRef]
21. Sodangi, M.; Kazmi, Z.A. Integrated evaluation of the impediments to the adoption of coconut palm wood as a sustainable material for building construction. *Sustainability* **2020**, *12*, 7676. [CrossRef]
22. Negro, E.; D'Amato, M.; Cardinale, N. Non-invasive Methods for Energy and Seismic Retrofit in Historical Building in Italy. *Front. Built Environ.* **2019**, *5*, 125. [CrossRef]
23. Gandage, A.S.; Rao, V.R.V.; Sivakumar, M.V.N.; Vasan, A.; Venu, M.; Yaswanth, A.B. Effect of Perlite on Thermal Conductivity of Self Compacting Concrete. *Procedia Soc. Behav. Sci.* **2013**, *104*, 188–197. [CrossRef]
24. Universal Technologies Concellmex S.A. DE C.V. Universal Technologies Concellmex Webpage. Available online: http://concellmex.com.mx/ (accessed on 17 November 2020).
25. Font, A.; Borrachero, M.V.; Soriano, L.; Monzó, J.; Mellado, A.; Payá, J. New eco-cellular concretes: Sustainable and energy-efficient materials. *Green Chem.* **2018**, *20*, 4684–4694. [CrossRef]
26. Bardenhagen, S.G.; Brydon, A.D.; Guilkey, J.E. Insight into the physics of foam densification via numerical simulation. *J. Mech. Phys. Solids* **2005**, *53*, 597–617. [CrossRef]
27. Chica, L.; Alzate, A. Cellular concrete review: New trends for application in construction. *Constr. Build. Mater.* **2019**, *200*, 637–647. [CrossRef]
28. Shafigh, P.; Asadi, I.; Akhiani, A.R.; Mahyuddin, N.B.; Hashemi, M. Thermal properties of cement mortar with different mix proportions. *Mater. Constr.* **2020**, *70*, 1–12. [CrossRef]
29. Prusty, J.K.; Patro, S.K.; Basarkar, S.S. Concrete using agro-waste as fine aggregate for sustainable built environment—A review. *Int. J. Sustain. Built Environ.* **2016**, *5*, 312–333. [CrossRef]
30. Eiras, J.N.; Segovia, F.; Borrachero, M.V.; Monzó, J.; Bonilla, M.; Payá, J. Physical and mechanical properties of foamed Portland cement composite containing crumb rubber from worn tires. *Mater. Des.* **2014**, *59*, 550–557. [CrossRef]
31. ASTM International. *ASTM C1693—11. Standard Specification for Autoclaved Aerated Concrete (AAC)*; ASTM International: West Conshohocken, PA, USA, 2017.
32. American Concrete Institute. *ACI 523.3R-14 Guide for Cellular Concretes above 50 lb/ft 3 (800 kg/m3)*; American Concrete Institute: Farmington Hills, MI, USA, 2014; p. 17.
33. Diario Oficial de la Federacion. *NOM-018-ENER-2011. Aislantes Térmicos Para Edificaciones. Características y Métodos de Prueba*; Secretaria de Energia: Mexico City, Mexico, 2011.
34. *NMX-C-460-ONNCCE-2009. Industria de la Construcción—Aislamiento Térmico—Valor "R" para las Envolventes de Vivienda por Zona Térmica para la República Mexicana—Especificaciones y Verificación*; ONNCCE: Mexico City, Mexico, 2009.
35. International Organization for Standardization. *ISO 10456:2007 Building Materials and Products—Hygrothermal Properties—Tabulated Design Values and Procedures for Determining Declared and Design Thermal Values*; International Organization for Standardization: Geneva, Switzerland, 2007; p. 25.
36. Diario Oficial de la Federacion. *NOM-020-ENER-2011. Envolvente de Edificios Para Uso Habitacional*; Secretaria de Energia: Mexico City, Mexico, 2011.
37. Diario Oficial de la Federacion. *NOM-008-ENER-2001. Eficiencia Energetica en Edificaciones, Envolvente de Edificios no Residenciales*; Secretaria de Energia: Mexico City, Mexico, 2001.
38. Oropeza-Perez, I.; Ostergaard, P.A. Energy saving potential of utilizing natural ventilation under warm conditions—A case study of Mexico. *Appl. Energy* **2014**, *130*, 20–32. [CrossRef]
39. Comisión Nacional para el Uso Eficiente de la Energía (CONUEE). *Estudio de Caracterización del Uso de Aire Acondicionado en Viviendas de Interés Social en México*; Secretaria de Energia: Mexico City, Mexico, 2016; p. 80.
40. de Buen Rodríguez, O.; Hernández-Pérez, F.; Navarrete, J.I. *Análisis de la Evolución del Consumo Eléctrico del Sector Residencial Entre 1982 y 2014 e Impactos de Ahorro de Energía por Políticas Públicas*; Secretaria de Energia: Mexico City, Mexico, 2016; p. 12.

41. Zhu, M.; Pan, Y.; Huang, Z.; Xu, P. An alternative method to predict future weather data for building energy demand simulation under global climate change. *Energy Build.* **2016**, *113*, 74–86. [CrossRef]
42. Huld, T.; Paietta, E.; Zangheri, P.; Pascua, I.P. Assembling typical meteorological year data sets for building energy performance using reanalysis and satellite-based data. *Atmosphere* **2018**, *9*, 53. [CrossRef]
43. Hosseini, M.; Bigtashi, A.; Lee, B. Generating future weather files under climate change scenarios to support building energy simulation—A machine learning approach. *Energy Build.* **2021**, *230*, 110543. [CrossRef]
44. Farah, S.; Whaley, D.; Saman, W.; Boland, J. Integrating climate change into meteorological weather data for building energy simulation. *Energy Build.* **2019**, *183*, 749–760. [CrossRef]
45. Pyrgou, A.; Castaldo, V.L.; Pisello, A.L.; Cotana, F.; Santamouris, M. Differentiating responses of weather files and local climate change to explain variations in building thermal-energy performance simulations. *Sol. Energy* **2017**, *153*, 224–237. [CrossRef]
46. Chakraborty, D.; Elzarka, H.; Bhatnagar, R. Generation of accurate weather files using a hybrid machine learning methodology for design and analysis of sustainable and resilient buildings. *Sustain. Cities Soc.* **2016**, *24*, 33–41. [CrossRef]
47. Hong, T.; Chang, W.; Lin, H. A Sensitivity Study of Building Performance Using 30-Year Actual Weather Data. In Proceedings of the 13th Conference of International Building Performance Simulation Association, Chambéry, France, 26–28 August 2013; pp. 1–8.
48. Taseska, V.; Markovska, N.; Callaway, J.M. Evaluation of climate change impacts on energy demand. *Energy* **2012**, *48*, 88–95. [CrossRef]
49. Clarke, L.; Eom, J.; Marten, E.H.; Horowitz, R.; Kyle, P.; Link, R.; Mignone, B.K.; Mundra, A.; Zhou, Y. Effects of long-term climate change on global building energy expenditures. *Energy Econ.* **2018**, *72*, 667–677. [CrossRef]
50. Cellura, M.; Guarino, F.; Longo, S.; Tumminia, G. Climate change and the building sector: Modelling and energy implications to an office building in southern Europe. *Energy Sustain. Dev.* **2018**, *45*, 46–65. [CrossRef]
51. Wang, H.; Chen, Q. Impact of climate change heating and cooling energy use in buildings in the United States. *Energy Build.* **2014**, *82*, 428–436. [CrossRef]
52. Resch, E.; Andresen, I.; Cherubini, F.; Brattebø, H. Estimating dynamic climate change effects of material use in buildings—Timing, uncertainty, and emission sources. *Build. Environ.* **2021**, *187*, 107399. [CrossRef]
53. Karimpour, M.; Belusko, M.; Xing, K.; Boland, J.; Bruno, F. Impact of climate change on the design of energy efficient residential building envelopes. *Energy Build.* **2015**, *87*, 142–154. [CrossRef]
54. Figueiredo, R.; Nunes, P.; Panão, M.J.N.O.; Brito, M.C. Country residential building stock electricity demand in future climate—Portuguese case study. *Energy Build.* **2020**, *209*, 109694. [CrossRef]
55. Wong, S.L.; Wan, K.K.W.; Li, D.H.W.; Lam, J.C. Impact of climate change on residential building envelope cooling loads in subtropical climates. *Energy Build.* **2010**, *42*, 2098–2103. [CrossRef]
56. Pérez-Andreu, V.; Aparicio-Fernández, C.; Martínez-Ibernón, A.; Vivancos, J.L. Impact of climate change on heating and cooling energy demand in a residential building in a Mediterranean climate. *Energy* **2018**, *165*, 63–74. [CrossRef]
57. Invidiata, A.; Ghisi, E. Impact of climate change on heating and cooling energy demand in houses in Brazil. *Energy Build.* **2016**, *130*, 20–32. [CrossRef]
58. Verichev, K.; Zamorano, M.; Carpio, M. Effects of climate change on variations in climatic zones and heating energy consumption of residential buildings in the southern Chile. *Energy Build.* **2020**, *215*, 109874. [CrossRef]
59. Flores-Larsen, S.; Filippín, C.; Barea, G. Impact of climate change on energy use and bioclimatic design of residential buildings in the 21st century in Argentina. *Energy Build.* **2019**, *184*, 216–229. [CrossRef]
60. Gercek, M.; Arsan, Z.D. Energy and environmental performance based decision support process for early design stages of residential buildings under climate change. *Sustain. Cities Soc.* **2019**, *48*, 101580. [CrossRef]
61. *NMX-C-077-ONNCCE-1997: Industria de la Construcción—Agregados para Concreto—Análisis Granulométrico—Método de Prueba*; ONNCCE: Mexico City, Mexico, 1997.
62. Borbon-Almada, A.; Rodriguez-Muñoz, N.; Najera-Trejo, M. Energy and Economic Impact on the Application of Low-Cost Lightweight Materials in Economic Housing Located in Dry Climates. *Sustainability* **2019**, *11*, 1586. [CrossRef]

63. *NMX-C-144-ONNCCE-2015: Industria de la Construcción—Cementantes Hidráulicos—Requisitos Para El Aparato Usado En la Determinación de la Fluidez de Morteros*; ONNCCE: Mexico City, Mexico, 2015.
64. NMX-C-061-ONNCCE-2015. *Industria de la Construcción—Cementantes Hidráulicos—Determinación de la Resistencia a la Compresión de Cementantes Hidráulicos*; ONNCCE: Mexico City, Mexico, 2015.
65. GACETA. *Oficial de la Ciudad de México Norma Técnica Complementaria para Mampostería. Reglamento de Construcción del Distrito Federal*; Gobierno de la Ciudad de Mexico: Mexico City, Mexico, 2020.
66. *NMX-C-486-ONNCCE-2014: Industria de la Construcción—Mampostería—Mortero para Uso Estructural—Especificaciones y Métodos de Ensayo*; ONNCCE: Mexico City, Mexico, 2014.
67. ASTM International. *ASTM C109/C109M. Standard Test Method for Compressive Strength of Hydraulic Cement Mortars (Using 2-in. or [50-mm] Cube Specimens)*; ASTM International: West Conshohocken, PA, USA, 2016; p. 10.
68. International Organization for Standardization. *ISO 8302:1991 Thermal Insulation—Determination Of Steady-State Thermal Resistance and Related Properties—Guarded Hot Plate Apparatus*; International Organization for Standardization: Geneva, Switzerland, 1991.
69. ASTM International. *ASTM C177—19: Standard Test Method for Steady-State Heat Flux Measurements and Thermal Transmission Properties by Means of the Guarded-Hot-Plate Apparatus*; ASTM International: West Conshohocken, PA, USA, 2019.
70. Comite Europeen de Normalisation. *EN 1946-2:1999. Thermal Performance of Building Products and Components—Specific Criteria for the Assessment of Laboratories Measuring Heat Transfer Properties—Part 2: Measurements by Guarded Hot Plate Method*; Comite Europeen de Normalisation: Brussels, Belgium, 1999.
71. Asociación Española de Normalización (UNE). *UNE-EN 12664:2002: Thermal Performance of Building Materials and Products. Determination of Thermal Resistance by Means of Guarded Hot Plate and Heat Flow Meter Methods. Dry and Moist Products of Medium and Low Thermal Resistance*; Asociación Española de Normalización (UNE): Madrid, Spain, 2002.
72. Asociación Española de Normalización (UNE). *UNE-EN 12667:2002. Thermal Performance of Building Materials and Products. Determination of Thermal Resistance by Means of Guarded Hot Plate and Heat Flow Meter Methods. Products of High and Medium Thermal Resistance*; Asociación Española de Normalización (UNE): Madrid, Spain, 2002.
73. Asociación Española de Normalización (UNE). *UNE-EN 12939:2001. Thermal Performance of Building Materials and Products—Determination of Thermal Resistance by Means of Guarded Hot Plate and Heat Flow Meter Methods—Thick Products of High and Medium Thermal Resistance*; Asociación Española de Normalización (UNE): Madrid, Spain, 2001.
74. Deutsches Institut für Normung. *DIN 52612-2:1984. Testing of Thermal Insulating Materials. Determination of Thermal Conductivity by Means of the Guarded Hot Plate Apparatus. Conversion of the Measured Values for Building Applications*; Deutsches Institut für Normung (DIN): Berlin, Germany, 1984.
75. ASTM International. *ASTM D5334 Standard Test Method for Determination of Thermal Conductivity of Soil and Soft Rock by Thermal Needle Probe Procedure 8*; ASTM International: West Conshohocken, PA, USA, 2014.
76. ASHRAE. *ASHRAE 55:2004 Thermal Environmental Conditions for Human Occupancy*; ASHRAE: Atlanta, GA, USA, 2004; p. 30.
77. Instituto Nacional de Estadística y Geografía Population Census. Available online: https://www.inegi.org.mx/programas/intercensal/2015/#Tabulados (accessed on 12 November 2020).
78. Secretaria de Energia & Banco Mundial. *Evaluación Rápida del Uso de la Energía*; Secretaria de Energia: Hermosillo, Sonora, Mexico, 2016.
79. Intergovernmental Panel on Climate Change (IPCC). *Special Report Emissions Scenarios: Summary for Policymakers*; United Nations Environment Program (UNEP): Nairobi, Kenya, 2000.
80. Intergovernmental Panel on Climate Change. *Special Report on Emission Scenarios*; Cambridge University Press: Cambridge, UK, 2000.
81. Ni, F.M.W.; Averyanov, S.; Melese, E.; Tighe, S. Characterization of lightweight cellular concrete. In Proceedings of the 7th International Material Speciaty Confenrence 2018, Held as Part of the Canadian Society Civil Engineering Annual Conference, Fredericton, NB, Canada, 13–18 June 2018; Volume 1923, pp. 45–55.

82. Mohd Sari, K.A.; Mohammed Sani, A.R. Applications of Foamed Lightweight Concrete. *MATEC Web Conf.* **2017**, *97*, 1–5. [CrossRef]
83. Comisión Reguladora de Energía. Gobierno de México Factor de emisión del Sistema Eléctrico Nacional 2019. Available online: https://www.gob.mx/cms/uploads/attachment/file/538473/Factor_emision_electrico_2019.pdf (accessed on 17 November 2020).

Article

Affordable, Energy-Efficient Housing Design for Chile: Achieving Passivhaus Standard with the Chilean State Housing Subsidy

Aner Martinez-Soto [1,*], Yarela Saldias-Lagos [1], Valentina Marincioni [2] and Emily Nix [2]

1 Department of Civil Engineering, Faculty of Engineering and Science, Universidad de La Frontera, Temuco 4780000, Chile; y.saldias01@ufromail.cl

2 UCL Institute for Environmental Design and Engineering, University College London, London WC1H 0NN, UK; v.marincioni@ucl.ac.uk (V.M.); emily.nix.12@ucl.ac.uk (E.N.)

* Correspondence: aner.martinez@ufrontera.cl; Tel.: +56-45-259-6816

Received: 12 July 2020; Accepted: 10 August 2020; Published: 22 October 2020

Abstract: In Chile, it is estimated that the energy demand will continue to increase if substantial energy efficiency measures in housing are not taken. These measures are generally associated with technical and mainly economic difficulties. This paper aims to show the technical and economic feasibility of achieving Passivhaus standard house in Chile, considering the budget of the maximum state subsidy currently available (Chilean Unidad de Fomento (CLF) 2000 ≈ 81,000 USD). The design was simulated in the Passive House Planning Package software to determine if the house could be certified with the selected standard. At the same time, the value of all the items was quantified in order not to exceed the stipulated maximum budget for a house considered as affordable. It was shown that in terms of design it is possible to implement the Passivhaus standard given the current housing subsidy. The designed housing ensures a reduction of 85% in heating demand and a 60% reduction in CO_2 emissions during the operation, compared to an average typical Chilean house.

Keywords: energy consumption; building construction; Passivhaus; affordable housing

1. Introduction

The residential sector is responsible for 40% of global carbon dioxide, thus energy efficiency measures in this sector would help to support emission reduction targets for climate mitigation [1,2]. In Chile, the residential sector accounts for 30% of the primary energy consumption [3,4] with the majority of energy consumption used for space heating [5]. A study reviewing a representative sample of homes in Chile found that the majority are energy inefficient and do not offer adequate thermal comfort [6]. A total of 67% of the total housing stock in the last four decades was built with a housing subsidy or as part of the programs of Basic and Progressive Housing [7]. The housing subsidy is a state monetary contribution that allows families to purchase or construct a house, with the subsidy ranging between ≈35,000 and 81,000 USD, depending on the family's economic circumstances and the size of the house you want to build or buy.

Several authors have identified that the thermal insulation in the majority of homes built under the subsidy programme is insufficient or simply non-existent [6,8]. As such, significant focus is needed to improve the energy efficiency across the residential sector in Chile. Improved dwelling design has shown to reduce energy consumption as well as providing safer and more comfortable indoor environments. Current studies recommend a combination of bio-climatic design features, such as optimizing the orientation, size of glazing, effective ventilation, solar shading [9–11], combined with passive heating and cooling [12,13]. New architectural approaches [14] and supporting regulations are vital for an improved energy efficiency across the residential sector. In this regard, many high-income countries have established building regulations which have requirements for energy efficiency [15–17]. In 2000, Chile established thermal regulations that consider the energy performance of residential houses and stipulate minimum levels of the insulation [18]. However, these regulations are likely to fall short of achieving energy efficiency savings necessary to meet emission reduction targets.

There are several established standards and evaluation frameworks for supporting the design and assessment of energy-efficient housing. Conceived in Germany, the Passivhaus standard [19] was the first standard developed to achieve indoor thermal comfort with low energy demand, while reducing the CO_2 emissions throughout the useful lifecycle of the home; the Passivhaus standard has been commonly used for housing and is now established in many countries [20]. Other standards include the Minergie P standard [21] in Switzerland, the CasaClima standard in Italy, and the Association for Environment Conscious Building (AECB) standard in the UK.

The Passivhaus standard was originally designed for cold climates, and focused on Central and Northern Europe [20], but it has been proven that is possible to design Passivhaus buildings for different climate zones of the world [22]. A case study in a tropical climate (Malaysia) shows that a reduction of the solar radiation heat gain by wall and roof insulation, combined with shading measures, led to a significant reduction in cooling load in a Passivhaus [23]. A study in Australia demonstrated, through a year of monitoring, that a Passivhaus required 64% less energy than an average comparable house in the same city (Chifley) [24]. An evaluation of the building performance in several climates the Southern Hemisphere demonstrated that the requirements of the Passivhaus standard were fulfilled for all Southern Hemisphere cases [25]; the study was based on 38 different locations (including Santiago de Chile). A study realized in Russia considering seven different climates (Monsoon-influenced warm-summer humid continental Dwb, Subarctic climate Dfc, Warm-summer humid continental climate DfB, Tundra climate ET in the Köppen climate classification) show that all of the case studies complied with the Passivhaus standard with regards to the primary energy demand [26]. In general, different studies [22–30] have proven that the Passivhaus standard is suitable for different climates. Thus, Passivhaus could provide a useful guide to support the development of energy-efficient housing in Chile. However most of the studies are focused on the technical aspects, benefits of thermal comfort, and energy performance while approaches based on the costs associated with the construction of a Passivhaus are scarce [31,32]. Some authors have mentioned that economic aspects represent the greatest challenge for the application of the Passivhaus standard [33,34] and this challenge must be addressed individually in each country because the costs are not comparable between countries.

A previous study assessed the barriers and opportunities associated with the implementation of energy efficiency standards in Chilean social housing and determined that their implementation would not be economically feasible [4]; the study found that building in compliance with the Passivhaus standard has a higher cost per square meter [4] than a conventional house (ca. 30%). This has also been found in other countries with limited resources [34–36]. However, in the current market there is a wide variety of increasingly affordable insulation materials with low thermal conductivity.

This work aimed to evaluate the technical and economic feasibility of a Passivhaus dwelling in Chile, considering the current level of state subsidy for housing construction. We developed a design for a passive house in Chile and evaluated its affordability. In cases where construction costs are unaffordable to families, it is of vital importance to know what construction standards can be achieved. In this sense, this study aims to show the possibility of building an affordable dwelling with low energy demand.

2. Materials and Methods

2.1. Location

According to MINVU (Ministerio de Vivienda y Urbanismo, Ministry of Housing and Urbanism) statistics, more than 35% of the housing subsidies are used in Santiago [37]. Additionally, Santiago is located in the climate zone with average temperatures and has the largest number of dwellings in Chile. Furthermore, between 2011 and 2016 of the 36,000 housing subsidies delivered through the subsidy programme (see Section 2.2) around 13,500 (≈38%) were delivered in the Metropolitan Region of Santiago. Therefore, the case study Santiago was selected as it better represents the national residential housing stock and it is expected to have the greatest concentration of subsidized housing.

Santiago can be classified as a Mediterranean climate (Csb) in the international Köppen climate classification. According to the Chilean Meteorological Office [38], the average minimum temperature in Santiago in the last decade was 7.9 °C (normally in June) and the average maximum temperature was 21.8 °C (always in January). The average solar global radiation fluctuates between 68 and 252 KWh/m^2·month [39], the average relative humidity is 74%, and air velocity is 4 m/s [25].

The Passive House Planning Package (PHPP) provides climate data for Santiago. In Figure 1, it can be seen that the outdoor temperature reaches its highest level in the months 6 (December in this study) to 8 (February in this study); the lowest temperature occurs in months 12 (June in this study), 1 and 2 (July and August in this study). Correspondingly, the highest level of solar radiation occurs in the hottest months.

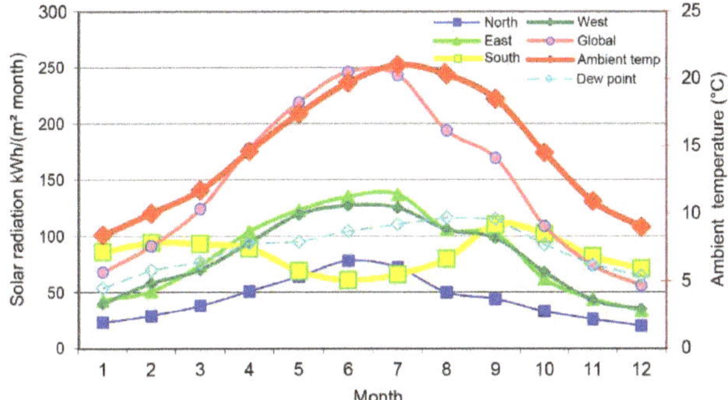

Figure 1. Solar radiation and outdoor temperature, monthly average. Climate Santiago, Chile. Note: For this study month 1 is July and month 7 is January. Data extracted from simulation in Passive House Planning Package (PHPP).

2.2. Affordability

In the last decade, the housing price index in Chile increased by 30% [40], while the median per capita income increased by 6% [41]. For this reason, the state subsidy in Chile is of great importance to support the purchase and construction of affordable housing. There are currently four subsidies for the purchase or construction of houses, which depend on the available budget and the degree of social vulnerability of the applicants. With subsidy DS 49 it is possible to buy or construct a house of up to 35,000 USD and with subsidy DS1 parts 1, 2, and 3 with which it is possible to buy or construct a house of up to 40,000 USD, 55,000 USD, and 81,000 USD respectively. Here it is important to note, that the construction of a traditional house in Chile costs on average 41,000 USD [42].

This paper presents the development of an affordable dwelling for a family in Chile, taking into consideration the highest subsidy (DS1) currently available for building a house. The subsidy DS1 offers a maximum budget of 2000 CLF/81,000 USD (CLF = Unidad de Fomento, CODE CLF according to ISO 4217 is a Unit of amount that is used in Chile adjusted with the inflation. A total of 2000 CLF is currently equal to approximately 81,000 USD). These resources were allocated towards a highly insulated thermal envelope and for the purchase of equipment required to meet the criteria of the Passivhaus Standard. This paper evaluates if the subsidy DS1 allows the construction of a Passivhaus dwelling with a floor area of 100 m^2.

2.3. Development of Design

The proposed dwelling was simulated with PHPP, the calculation tool used to evaluate the compliance of designs with the Passivhaus standard. At the same time, a financial plan for the construction of the house was calculated, considering all the costs associated with the construction in order not to exceed the maximum available budget determined by the subsidy. Within the design stage, the following aspects were considered:

(a) Architecture

For the architectural design of the dwelling (Figure 2), a compact, single-level house was considered. A certified Passivhaus professional using the principles of bioclimatic design developed the concept. The compact rectangular shape reduces the thermal energy losses by transmission because of the smaller number of surfaces exposed to the outdoor environment and by minimizing the number of thermal bridges. The majority of glazing is found on the north façade, as this will provide solar gains for heating that will be key for a low energy demand in winter. This is because Santiago is in the southern hemisphere and therefore the north façade provides most solar gains. The opposite case occurs in the Northern hemisphere where preferable planning window is in the south façade for solar gains.

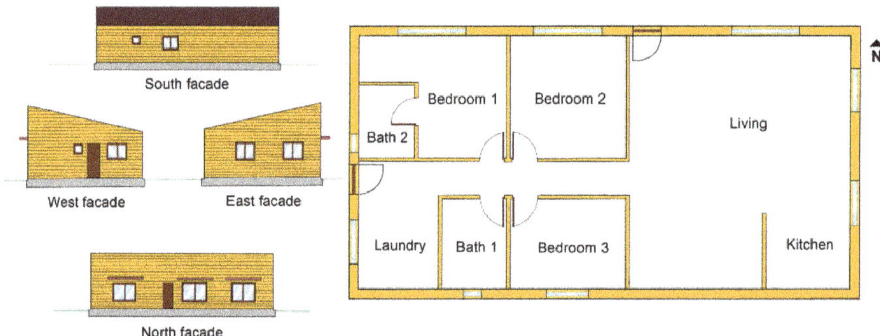

Figure 2. The architectural design of the house. Note: For data entry into PHPP, the cardinal direction "North" was changed by 180°. This is the recommended approach for buildings in the Southern hemisphere as per the PHPP User's Manual.

The designed dwelling consists of the same number of rooms as a conventional three-bedroom house, with two bathrooms, a kitchen, a living-dining room, and a space for laundry and a desk. Most living areas, such as the living room and bedrooms, were located on the north side to maximize the occupants' exposure to sunlight, leaving the services on the south side. Additionally, the number of internal divisions was minimized to reduce the junctions between internal partitions and perimeter walls and to avoid possible thermal bridges. The plot area is 112 m^2 while the useful floor area is 100 m^2, as the thickness of the perimeter walls covers about 12 m^2 of the plot area.

(b) Thermal envelope

For the thermal envelope, the design aimed to minimize the energy losses, with a particular focus on reducing thermal bridges. The external walls were composed of two elements separated with pieces of impregnated pressure-vacuum (IPV) timber; the design includes two 15 mm-thick Oriented Strand Boards, which support the load, hence avoiding the use of crosses and diagonals that would increase heat transmission (Figure 3). The system designed for the external walls allows having an envelope with continuous thermal insulation and with uniform thickness throughout its length, even in the corners of the wall, with the aim of minimizing the thermal bridges (seeing Appendix A). The insulating material corresponds to 120 mm-thick mineral wool, of density 40 kg/m^3 and a thermal conductivity of 0.042 W/(mK). Using the PHPP spreadsheets, it was determined that the thermal transmittance of the perimeter walls is 0.244 W/(m^{2}K). This represents a much lower value than what is required by the Chilean building regulations (Urban General Urban Planning and Construction Ordinance [18]), and even lower than that required in the areas where additional requirements related to thermal transmittance are applied due to the Atmospheric Decontamination Plan for particulate matter [43] (see the example for Temuco and Padre Las Casas in Table 1). It is noted as a disadvantage that the total thickness of the perimeter wall is 24.5 cm, so that multiplied by the perimeter of the house results in an area of approx. 12 m^2, which must be removed from the living area of the house. In Chile, a typical thickness of a wall is 14 cm [5].

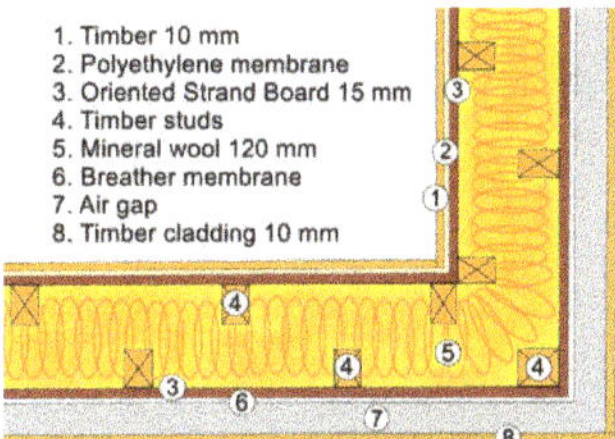

Figure 3. Detailed perimeter wall, plan view. Own elaboration.

Table 1. Requirements of transmittance and thermal resistance for elements of the envelope according to Article 4.1.10 General Urban Planning and Construction Ordinance and according to the Atmospheric Decontamination Plan for the communes of Temuco and Padre Las Casas in Chile.

Zone	Roofing		Walls		Floors	
	U	R_{TOT}	U	R_{TOT}	U	R_{TOT}
1	0.84	1.19	4.00	0.25	3.60	0.28
2	0.60	1.67	3.00	0.33	0.87	1.15
3	0.47	2.13	1.90	0.53	0.70	1.43
4	0.38	2.63	1.70	0.59	0.60	1.67
5	0.33	3.03	1.60	0.63	0.50	2.00
6	0.28	3.57	1.10	0.91	0.39	2.56
7	0.25	4.00	0.60	1.67	0.32	3.13
PDA Temuco, Padre las Casas	0.28	3.57	0.45	2.22	0.50	2.00

Thermal transmittance values U in W/m^{2}K and total thermal resistance values in m^{2}K/W. The proposed PassivHaus is located in zone 3.

The ceiling was considered timber-framed with 50 × 50 mm joists, 120 mm mineral wool thermal insulation, and Oriented Strand Board (OSB) on the inside. As in the walls, the thermal insulation layer is available in a continuous and uniform thickness (Figure 4). To calculate the thermal transmittance, only the ceiling structure and the insulating layer are considered, and the attic space and roof are not considered, due to their little influence on the calculations as recommended by the PHPP manual. The thermal transmittance obtained for the ceiling is 0.238 (W/m^{2}K).

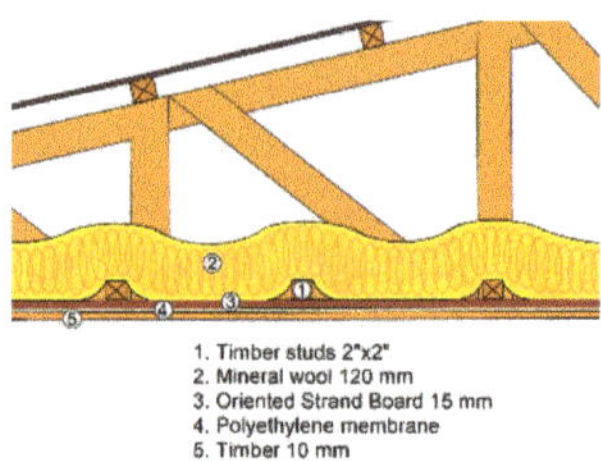

Figure 4. Ceiling cross-section, view in section. Own elaboration.

The construction for the floor (Figure 5) consists of expanded polystyrene of high density (20 kg/m^3) in a thickness of 100 mm on a screed of 300 mm and on it a solid floor layer of 100 mm of lightweight concrete with low thermal conductivity. To break the thermal bridge that occurs in the wall–floor junction, a layer of additional insulation (polyurethane in a thickness of 40 mm and height of 100 mm) is placed between the footplate and the wall. With this, it was determined that the linear thermal transmittance value at that point is negative (see Appendix A). The concrete of the foundation and sole plate are based on light aggregates of expanded clay, which has a thermal conductivity of 0.330 W/(mK). With this construction, a value of 0.307 W/m^{2}K was obtained for the thermal transmittance of the floor.

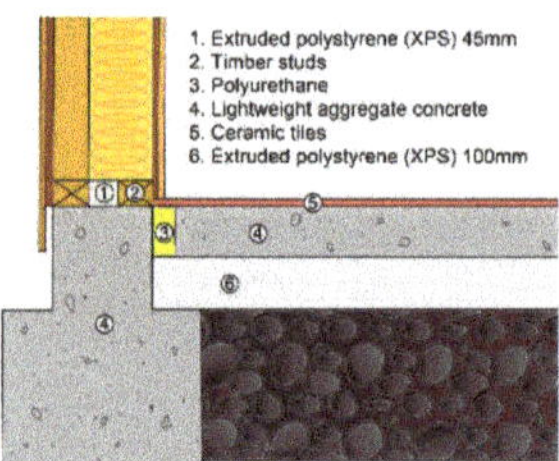

Figure 5. Details of the wall–floor junction, view in section. Own elaboration.

The climate classification used in Table 1, is according to the Chilean national Norm Ordenanza General de Urbanismo y Construcciones (General Urban Planning and Construction Ordinance), which establishes thermal zoning for new housing designs located in different regions of the country. According to the Instituto Nacional de Estadisticas of Chile [44], it is possible to link the national climate classification to the international Köppen climate classification. For example, the north (Zone 1) and south of Chile (Zone 7) can be classified as Arid desert climate (BWk) and Oceanic climate (Cfc), respectively. In this sense, the location considered in this work (Santiago, Zone 3) can be classified as a Mediterranean climate (Csb).

(c) Thermal bridges

The PHPP recommends the identification and classification of thermal bridges. In addition, to determine the energy demand, the linear transmittance with its respective length in meters must be determined. In this work the coefficients of linear transmittance were determined with the THERM software, in accordance with ISO 10211:2007. The results are presented in Appendix A. Here we sought to minimize all the geometric thermal bridges.

(d) Windows and shading

To optimize passive solar gains, it is essential to know the predominant elevation angle of the sun in winter and summer. Figure 6 shows the predominant elevation of the sun in the city of Santiago (33° in winter and 79° in summer). The design of the house considered an eave of 50 cm and an overhang on the north-facing window of 30 cm. In summer, these measures protect a large part of the north wall from direct solar radiation; in winter, all the north-facing windows receive direct solar radiation. It considered eaves only in the north façade, because the radiation does not arrive with such intensity in the east and west façades, as it is present only in the early hours of the morning and the late hours of the afternoon where the temperature is on average 13 °C [45]. The south façade is not exposed to direct radiation throughout the day; it only receives diffuse and reflected radiation, which has a lower influence on energy use. In the north-facing windows, solar gains were 972 kWh/yr, while transmission losses reached 698 kWh/yr.

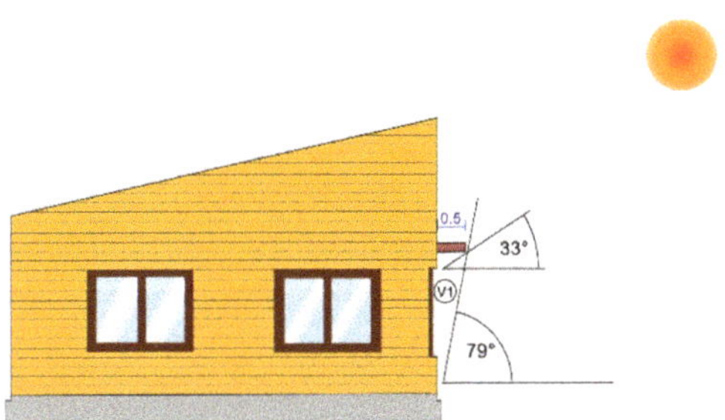

Figure 6. Shading produced by eaves on the north façade of the building. Own elaboration.

The windows considered in this work correspond to double glazing with timber frame with a U-value = 1.8 W/m^{2}K and g = 0.74 which is currently available in the Chilean market [46]. This type of windows has a transmittance that is 38% lower than that established as the limit for traditional windows (2.9 W/m^{2}K) in current Chilean regulations [47].

(e) Ventilation with heat recovery

The value of the total volume enclosed in the house designed is 245 m^3, so to ensure the air renewal required by the Passivhaus standard was considered an ECOWATT series fan, with heat recovery, high efficiency (between 86% and 92%), with a maximum air renewal flow rate of 325 m^3/h and consumption of less than 40 W. The building is heated with a Compact Heat Pump System (Aerex—PHK 180). For this heat pump, the effective heat recovery efficiency is 80% and useful air flow rates are 130–230 m^3/h with a COP_{heat} between 2.85 and 3.31.

2.4. Simulation Method

The designed house was simulated and evaluated with the calculation software PHPP (Passivhaus Planning Program) version 8.5, which is an extensively tested and validated simulation tool and is used to verify the fulfilment of the requirements for Passivhaus certification.

Although the Passivhaus standard establishes fixed energy demand limits for building design, this is rather a general concept aimed at achieving thermal comfort by heating or cooling an air volume with guaranteed quality [48]. In this sense, the design and execution of homes with Passivhaus standard have an unlimited number of constructive solutions; however, the following requirements must be met:

(a) Maximum annual energy demand for heating or cooling: 15 kWh/m^2.
(b) Annual total energy consumption for all systems (heating, cooling, hot water, electricity) not exceeding 120 kWh/m^2.
(c) Test of air tightness at n50 not higher than 0.6/h, value obtained by the "Blower Door" test.
(d) Interior surface temperatures of the thermal envelope during winter >17 °C and <25 °C during summer.

This study also considers the indoor set-point temperatures to be 20 and 25 °C in winter and summer, respectively.

The certification of a house with Passivhaus standard can be obtained through the PassivHaus Institut, attaching the PHPP spreadsheets, executive project documentation, technical information, air tightness test according to UNE EN 13829 standard, balance protocol ventilation system, inspection, and photographic documentation of the work. This paper considers the design stage, where PHPP is used to design a building that complies to points a, b, and d.

2.5. Costing of Design

The design of the house presented in this paper did not only consider meeting the criteria established for a Passivhaus but also a limited budget within what is regarded as affordable housing (construction budget not greater than CLF 2000 ≈ 81,000 USD). Here, CLF was used for two reasons: (a) the amount of the subsidy is delivered in that currency; (b) the value of CLF is adjusted monthly based on the Consumer Price Index (CPI), which is used to measures changes in the price level of the market of consumer goods and services purchased. The CLF is the principal measure to determine the real values of housing and any secured loan (monetary item), either private or of the Chilean government.

The calculation of the cost of the proposed house is carried out under a Unit Price Analysis based on the software ONDAC [49]. Here the price for workers, materials, transport, and tax is considered, these are based on the current market rates for construction in Chile.

The Unit price in ONDAC as well the real value of the subsidy are monthly actualized. This implies that, if the costs vary in the construction phase of housing in the future, the budget would be adjusted and covered by the amount delivered in the subsidy.

3. Results

Based on the PHPP tool under the climate conditions for the case study proposed here, energy is required only for heating. The summary of the results obtained from the PHPP (Appendix C) show that both the energy demand in heating, the heating load, and the percentage of overheating time are below the established requirements of the standard. (Table 2).

Table 2. Results and verification of compliance requirements Passivhaus (see Appendix C).

Parameter	Results	Passivhaus Requirement	Standard Compliance Passivhaus
Annual heating demand	10 kWh/(m^2)	≤15 kWh/(m^{2}a)	Yes
Heating load	8 W/m^2	≤10 W/m^2	Yes
Thermal bridges ΔU	0.1 W/(m^{2}K)	-	-
Overheating frequency	7.8%	≤10%	Yes
Annual total energy demand	120 KWh/(m^2)	≤120 kWh/(m^{2}a)	Yes

Implementing Passivhaus standard as affordable housing, specifically in Santiago, would result in a significant reduction in energy demand and CO_2 emissions. However, to demonstrate that the standard can be implemented in colder climates, a simulation of the same house was also carried out in the Temuco climate, where the ambient temperature is on average 3.3 °C lower. The results show that the same house that has an annual energy demand of 10 kWh/m^2 for heating in the city of Santiago, whereas in Temuco it has a demand of 24 kWh/m^2. To comply with the annual heating energy demand limit (15 kWh/m^2), the surface area of the windows in the north façade was increased by 7.8 m^2 and an additional 10 mm thickness was added to the outer wall covering of the perimeter wall. The cost of this change means an initial budget increase in CLF 75 to a total CLF 1848, which does not exceed the budget limit (CLF 2000 ≈ 81,000 USD).

In Table 3, it is observed that the thermal transmittance of the walls of the designed housing represents one-eighth of the thermal transmittance required by the norm, while for the roof this was reduced by half. It is further observed that the regulations establish a maximum value of glazed surface (60%) without breaking it down according to geographical orientation.

Table 3. Comparison between conventional housing characteristics and the designed Passivhaus.

	Conventional Housing Values Based on [18,48]	**Passivhaus Housing**
	Value U (W/m^2K)	
Walls	1.9	0.244
Roof	0.47	0.238
Solid floor	Does not apply *	0.307
	% Window surface	
Total (DGU $2.4 < U < 3.6$ W/m^2K)	60% max	14%
North facade	Does not apply **	26%
Air changes/h	8/h	0.6/h

* The regulation does not consider values of thermal transmittance for floor in contact with the ground, only delivers values for ventilated floors, for thermal zone 3, for example, consider a thermal transmittance value of 0.7 W/m^2K. ** In the regulation surface of windows oriented towards the north is not considered, considers percentages in relation to the total surface of vertical walls.

For the complete execution of the house, a total budget of CLF 1773 (≈80,000 USD) was estimated (Appendix B). The per square-meter cost of the designed house is CLF 15.83 (approximately 640 USD/m^2). A conventional house can be a maximum of 140 m^2 with a total budget of CLF 2000 (≈81,000 USD). If the proposed house had a maximum area (140 m^2) the total cost would be CLF 2,216 (≈90,000 USD) and thus 10% higher than the maximum allowable subsidy. Therefore, achieving Passivhaus, while affordable given the current subsidy, may result in some restrictions to dwelling sizes particularly in the colder climate zones of Chile. Nevertheless, it is expected that the construction of affordable housing with the Passivhaus standard could represent a significant potential for energy savings in the residential sector.

4. Discussion

This study demonstrated the possibility of building a house considered as affordable in Chile with a budget less than CLF 2000 (≈81,000 USD) with the Passivhaus standard. It should be clarified that this study only considered costs relating to labor and material cost during the construction phase. The additional costs of managing, designing, or additional checks/testing were not considered due to the scarcity of data related to these aspects in the Chilean market. Consequently, it is necessary to focus efforts on gathering data to determine these additional potential costs related to the design and checks to achieve Passivhaus house. As awareness of Passivhaus increases within Chile and becomes more commonly practiced, it is likely that these costs will not significantly differ from current practices.

Another important aspect to discuss is the designed dwelling energy performance and its comparison with existing homes. Hatt et al. (2012) demonstrated that a standard house based on the Thermal Normative in Chile requires at least 86 kWh/m^2a for heating energy consumption [6]. If we consider a 100 m^2 house located in Santiago with a referential annual energy demand of 86 kWh/m^2a, the proposed Passivhaus represents just 11% of the standard house energy demand. The proposed design complies with the allowable Passivhaus energy consumption with thermal transmittance values (Table 3) that are likely too high for cold climates [22]. This has been similarly demonstrated by other authors that have shown that in warmer climates it is possible to reach the passive standard with higher values of thermal transmittance compared to cold climates [6,24].

Although it was possible to demonstrate the construction of Passivhaus within the current state subsidy, we found that there is likely to be a restriction in dwelling size. This restriction is likely to be more pronounced in cold climate regions of Chile, where average temperatures are much lower and may lead to issues of overcrowding if subsidy costs cannot be increased for these regions. For the construction of a traditional house compared one that complies with Passivhaus standard, a substantial difference in construction costs are noted (Appendix B). The initial cost of the Passivhaus is more than 40% higher than that of traditional Chilean housing in the construction phase. This is in line with similar work who found that achieving compliance with Passivhaus standard has around 30% higher cost per square meter [4]. In this study, the city of Santiago with a Mediterranean climate (Csb) was considered. It is expected that in the south of Chile, which experiences colder climates, greater investment will be needed to reach Passivhaus standard. Further work is needed to assess the viability of achieving energy efficiency housing given the available subsidy across all climate zones in Chile.

The annual energy cost of Passivhaus housing is cheaper compared to that of traditional housing, which leads to a lower accumulated cost of the Passivhaus over the years. Figure 7 shows the comparison of the accumulated cost (initial + operational) over 30 years for a Passivhaus and a traditional housing for the Chilean case (Appendix B). Both houses have the same floor area (100 m^2), for the Passivhaus a heating energy consumption of 10 kWh/m^2a was considered (Table 2) and for a traditional house an average heating energy consumption of 120 kWh/m^2a based on CDT, 2010 [5]. For both cases, electricity was considered as an energy source with a price of 0.18 USD/kWh. Approximately, in year 17 the accumulated costs are equivalent. After this period, the Passivhaus is economically more profitable, having 20% less of the accumulated cost for the year 30.

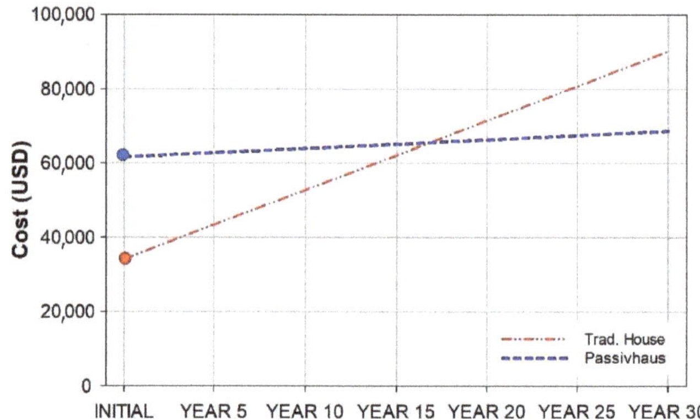

Figure 7. Accumulated annual energy cost of traditional house and Passivhaus. For the initial time, only the construction costs are considered (Appendix B).

5. Conclusions

In this paper, a house with the Passivhaus standard was designed with a budget of less than CLF 2000 (≈81,000 USD) in order to frame it within the subsidy program for affordable housing in Chile. The simulation of the housing design was made with the PHPP calculation software, which contains the requirements of the applied standard and finally defines if the house could be certified.

The construction materials proposed have a low thermal transmittance of the elements of the envelope. For example, in perimeter walls, a value of 0.244 W/m^2K was obtained, which represents a transmittance eight times less than what is required by regulation in the climate zone where the house was designed (Santiago). In addition, construction solutions were developed aimed at avoiding geometric thermal bridges and optimizing solar gains utilizing the appropriate percentage of the surface area of facing north windows.

It was demonstrated that in terms of design and construction costs, it is possible to develop an achieve Passivhaus standard of 112 m^2 within the Chilean state Subsidy of CLF 1773 (≈80,000 USD) to build a house. It was determined that the value of building with this standard reaches 640 USD per square meter. Although it was possible to achieve Passivhaus standard, it is likely to lead to a restriction in the floor area of a dwelling particularly in colder climate zones of Chile. Nevertheless, implementing Passivhaus standard would likely achieve a reduction in energy demand of 85% compared to a typical house in Chile (≈120 KWh/m^2). This is not only important for the economy of the individual families from weaker socio-economic groups but also it would help to significantly reduce energy consumption in Chile's residential sector.

Author Contributions: Conceptualization, A.M.-S.; Methodology, A.M.-S. and E.N.; Research, A.M.-S. and Y.S.-L.; Writing Y.S.-L. and V.M.; Review and Editing, V.M. and E.N. All authors have read and agreed to the published version of the manuscript.

Funding: This research received no external funding.

Conflicts of Interest: The authors declare no conflict of interest.

Appendix A

Table A1. Thermal Bridges.

	Identification of the Thermal Bridge	THERM of Simulation	Linear Thermal Transmittance (W/mK)
Vertical	1. **Corner of Wall** 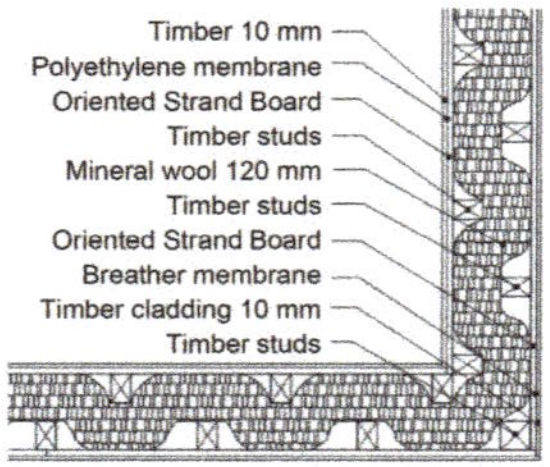 Timber 10 mm Polyethylene membrane Oriented Strand Board Timber studs Mineral wool 120 mm Timber studs Oriented Strand Board Breather membrane Timber cladding 10 mm Timber studs	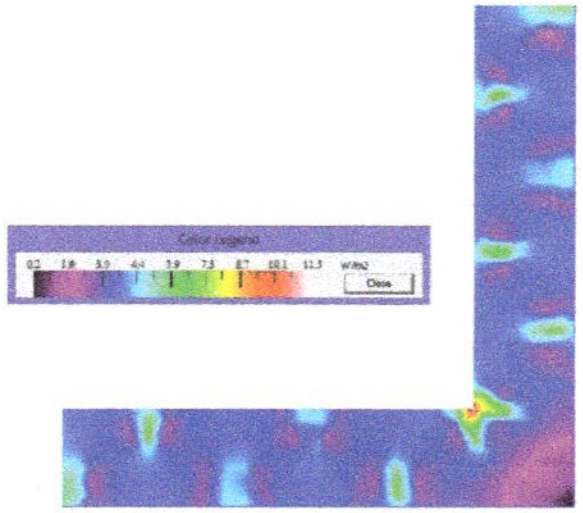	$\Psi = -0.164$
	1. Result: when simulating the thermal bridge with the THERM software and calculating the linear thermal transmittance, a negative value is obtained, therefore, it can be verified that the thermal bridge was broken and there was even a point improvement in the thermal envelope, allowing to have less energy flow at that point.		
Horizontal	2. **Wall and roof** 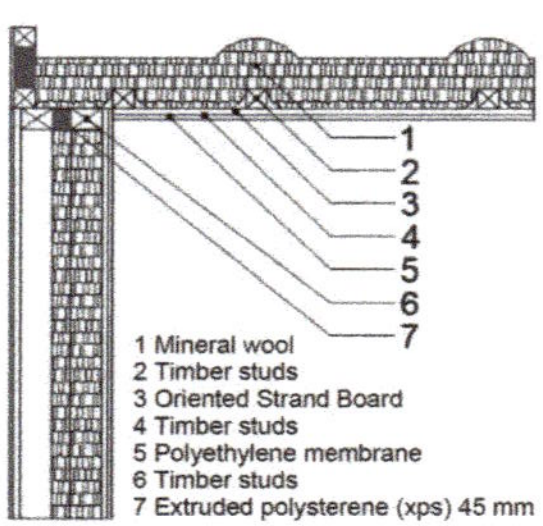 1 Mineral wool 2 Timber studs 3 Oriented Strand Board 4 Timber studs 5 Polyethylene membrane 6 Timber studs 7 Extruded polystyrene (xps) 45 mm	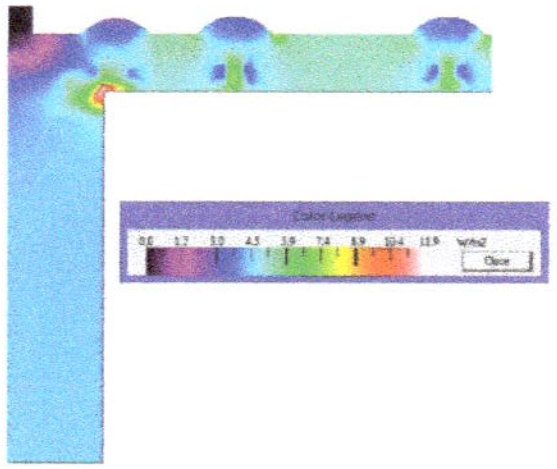	$\Psi = -0.093$
	2. Result: the simulation of the thermal bridge in the wall–sky encounter gives a negative linear thermal transmittance value, it can be said then that the thermal bridge was broken and there was a small point improvement in the thermal envelope.		

Table A1. *Cont.*

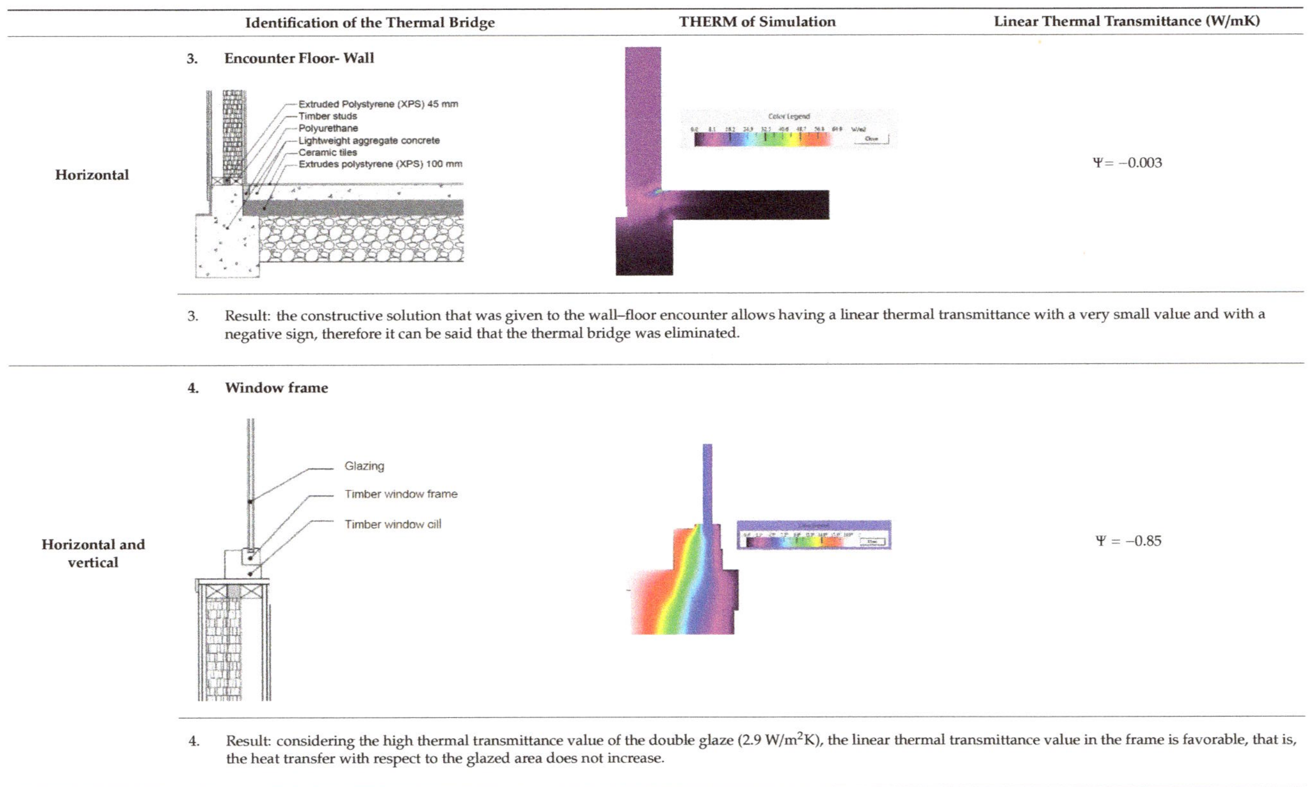

	Identification of the Thermal Bridge	THERM of Simulation	Linear Thermal Transmittance (W/mK)
Horizontal	3. **Encounter Floor- Wall**		Ψ = −0.003
	3. Result: the constructive solution that was given to the wall–floor encounter allows having a linear thermal transmittance with a very small value and with a negative sign, therefore it can be said that the thermal bridge was eliminated.		
Horizontal and vertical	4. **Window frame**		Ψ = −0.85
	4. Result: considering the high thermal transmittance value of the double glaze (2.9 W/m^{2}K), the linear thermal transmittance value in the frame is favorable, that is, the heat transfer with respect to the glazed area does not increase.		

Table A1. *Cont.*

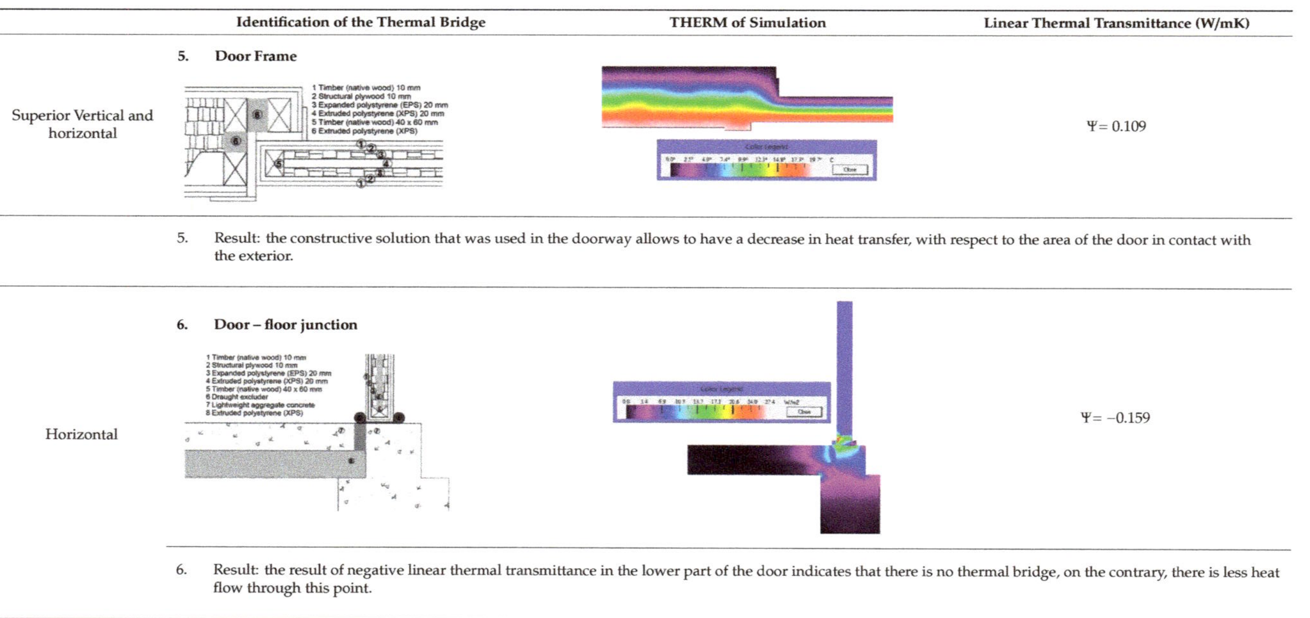

	Identification of the Thermal Bridge	THERM of Simulation	Linear Thermal Transmittance (W/mK)
Superior Vertical and horizontal	5. **Door Frame**		Ψ= 0.109
	5. Result: the constructive solution that was used in the doorway allows to have a decrease in heat transfer, with respect to the area of the door in contact with the exterior.		
Horizontal	6. **Door – floor junction**		Ψ= −0.159
	6. Result: the result of negative linear thermal transmittance in the lower part of the door indicates that there is no thermal bridge, on the contrary, there is less heat flow through this point.		

Appendix B

Table A2. Detailed Budget. Passivhaus.

Item	Description	Unit	Quantity	Unit Price (CLP)	Total (CLP)
1	Heavy work				
1.1	Scarp, levelling, and tracing	m^2	120	$3041	$364,889
1.2	Excavations	m^3	15	$8849	$132,739
1.3	Concrete				
1.3.1	Embedded H-5	m^3	2	$50,751	$101,501
1.3.2	Lightweight concrete foundation	m^3	17	$65,736	$1,117,504
1.3.3	Radier lightweight concrete	m^3	10	$63,306	$633,056
1.4	Structure of walls				
1.4.1	Perimeter walls	m^2	95	$23,123	$2,196,730
1.4.2	Internal partitions	m^2	75	$8.341	$625,573
1.5	Roof structure				
1.5.1	Trusses	m^2	116	$9027	$1,047,184
1.5.2	Wood structure	m^2	65	$12,553	$815,943
1.5.3	Ceiling curb	m^2	100	$12,391	$1,239,098
2	Thermal insulation				
2.1	Mineral wool 120 mm walls	m^2	95	$13,049	$1,239,665
2.2	Mineral wool 120 mm ceiling	m^2	100	$12,239	$1,223,910
2.3	Polyurethane exp. 100 mm floor	m^2	100	$8007	$800,685
2.4	Polyurethane 40 mm floor	Gl	1	$43,619	$43,619
3	Coatings				
3.1	Interiors				
3.1.1	Perimeter walls	m^2	105	$8000	$840,039

Table A2. *Cont.*

Item	Description	Unit	Quantity	Unit Price (CLP)	Total (CLP)
3.1.2	Dry zone partitions	m^2	110	$8000	$880,041
3.1.3	Wet zone partitions	m^2	40	$20,291	$811,620
3.2	Ceilings				
3.2.1	Dry zone	m^2	84	$8000	$672,031
3.2.2	Wet zone	m^2	16	$9782	$156,518
3.3	Exteriors				
3.3.1	Perimeter walls	m^2	120	$8000	$960,044
3.3.2	Eaves	m^2	5	$14,953	$74,763
3.3.4	Corrugated zinc cover 0.35	m^2	110	$7942	$873,626
3.3.5	Tinsmiths	Gl	1	$76,262	$76,262
4	Terminations				
4.1	Doors				
4.1.1	Interior	Un	5	$24,972	$124,862
4.1.2	Exterior	Un	2	$71,952	$143,905
4.2	Window DGU				
4.2.1	Window type 1 (2.0 × 1.5 m)	Un	3	$613,454	$1,840,361
4.2.2	Window type 2 (1.2 × 1.0 m)	Un	2	$306.285	$612,571
4.2.3	Window type 3 (0.5 × 0.5 m)	Un	2	$126,465	$252,931
4.3	Smock	Ml	105	$2641	$277,348
4.4	Cornices	Ml	105	$2479	$260,338
4.5	Interior ceramic pavement	m^2	100	$12,304	$1,230,390

Table A2. *Cont.*

Item	Description	Unit	Quantity	Unit Price (CLP)	Total (CLP)
4.6	Paint				
4.6.1	Exterior	m^2	170	$2552	$433,755
4.6.2	Interior	m^2	95	$2552	$242,393
5	Artefacts				
5.1	W.C.	Un	2	$70,571	$141,143
5.2	Sink with pedestal	Un	2	$58,421	$116,843
5.3	Bathtub	Un	2	$97,099	$194,198
5.4	Dishwasher with furniture	Un	1	$117,551	$117,551
5.5	Heater	Un	1	$149,951	$149,951
6	Equipment				
6.1	Fan with rec. of heat	un	1	$4,050,000	$4,050,000
7	Services and infrastructure				
7.1	Drinking water and sewage	Gl	1	$1,620,000	$1,620,000
7.2	Electricity	Gl	1	$1,000,000	$1,000,000
7.3	Gas	Gl	1	$810,000	$810,000
				Direct cost	$30,545,574
				Genera expenses 10%	$3,054,557
				Utility 20%	$6,109,115
				Total net	$39,709,246
				IVA 19%	$7,544,757
				Total cost	$47,254,003
				Total cost * CLF	1773

* Value of CLF July 2019.

Table A3. Detailed Budget. Traditional house.

Item	Description	Unit	Quantity	Unit Price (CLP)	Total (CLP)
1	Heavy work	—			
1.1	Scarp, levelling, and tracing	m^2	120	$878	$105,365
1.2	Excavations	m^3	15	$10,671	$160,060
1.3	Concrete				
1.3.1	Embedded H-5	m^3	2	$52,406	$104,812
1.3.2	Lightweight concrete foundation	m^3	17	$52,906	$899,409
1.3.3	Radier lightweight concrete	m^3	10	$76,613	$763,268
1.3.4	Reinforced concrete slab	m^3	5	$250,895	$1,254,475
1.4	Structure of walls				
1.4.1	Perimeter walls	m^2	95	$16,844	$1,600,255
1.4.2	Internal partitions	m^2	75	$3785	$283,883
1.5	Roof structure				
1.5.1	Trusses/wood structure	m^2	116	$10,488	$1,216,604
1.5.2	Ceiling curb	m^2	100	$2120	$212,024
2	Thermal insulation				
2.1	Mineral wool 90 mm walls	m^2	95	$3246	$308,360
2.2	Mineral wool 120 mm ceiling	m^2	100	$2139	$213,926
3	Coatings				
3.1	Interiors				
3.1.1	Perimeter walls	m^2	105	$3907	$410,208
3.1.2	Dry zone partitions	m^2	110	$1681	$184,930
3.1.3	Wet zone partitions	m^2	40	$8197	$327,873
3.2	Ceilings				

Table A3. *Cont.*

Item	Description	Unit	Quantity	Unit Price (CLP)	Total (CLP)
3.2.1	Dry zone	m^2	84	$2390	$200,782
3.2.2	Wet zone	m^2	16	$3308	$52,921
3.3	Exteriors				
3.3.1	Perimeter walls	m^2	120	$8867	$1,064,028
3.3.2	Eaves	m^2	5	$19,358	$96,788
3.3.3	Corrugated zinc cover 0.35	m^2	110	$7348	$808,246
3.3.4	Tinsmiths	Gl	1	$149,499	$149,499
4	Terminations				
4.1	Doors				
4.1.1	Interior	Un	5	$48,878	$244,389
4.1.2	Exterior	Un	2	$90,098	$180,195
4.2	Window DGU				
4.2.1	Window type 1 (2.0 × 1.5m)	Un	3	$145,309	$435,928
4.2.2	Window type 2 (1.2 × 1.0m)	Un	2	$137,058	$274,116
4.2.3	Window type 3 (0.5 × 0.5m)	Un	2	$59,858	$119,716
4.3	Smock	Ml	105	$2207	$231,770
4.4	Interior ceramic pavement	m^2	100	$13,027	$1,302,676
4.6	Paint				
4.6.1	Exterior	m^2	170	$4105	$697,777
4.6.2	Interior	m^2	95	$2903	$275,737
5	Artefacts				
5.1	W.C.	Un	2	$54,925	$109,850
5.2	Sink with pedestal	Un	2	$95,345	$190,690

Table A3. *Cont.*

Item	Description	Unit	Quantity	Unit Price (CLP)	Total (CLP)
5.3	Bathtub	Un	2	$65,605	$131,210
5.4	Dishwasher with furniture	Un	1	$62,180	$62,180
5.5	Heater	Un	1	$107,625	$107,625
6	Equipment				
6.1	Stove	Un	1	$360,783	$360,783
7	Services and infrastructure				
7.1	Drinking water and sewage	Gl	1	$253,281	$253,281
7.2	Electricity	Gl	1	$624,366	$624,366
7.3	Gas	Gl	1	$268,393	$268,393
				Direct cost	$16,020,018
				Genera expenses 10%	$1,602,002
				Utility 20%	$3,204,004
				Total net	$20,826,023
				IVA 19%	$3,956,944
				Total cost	$24,782,967
				Total cost * CLF	867

* Value of CLF July 2019.

Appendix C

Most important calculation sheets from PHPP.

				Requirements	Fulfilled?*
	Treated floor area	112,0	m²		
Space heating	Heating demand	**10**	**kWh/(m²a)**	15 kWh/(m²a)	**yes**
	Heating load	**8**	**W/m²**	10 W/m²	**yes**
Space cooling	Overall specif. space cooling demand		**kWh/(m²a)**	-	-
	Cooling load		**W/m²**	-	-
	Frequency of overheating (> 25 °C)	7,8	%	-	-
Primary energy	Heating, cooling, dehumidification, DHW, auxiliary electricity, lighting, electrical appliances	**119**	**kWh/(m²a)**	120 kWh/(m²a)	**yes**
	DHW, space heating and auxiliary electricity	49	kWh/(m²a)	-	-
	Specific primary energy reduction through solar electricity		kWh/(m²a)	-	-
Airtightness	Pressurization test result n_{50}	**0,6**	**1/h**	0,6 1/h	**yes**

Figure A1. Verification.

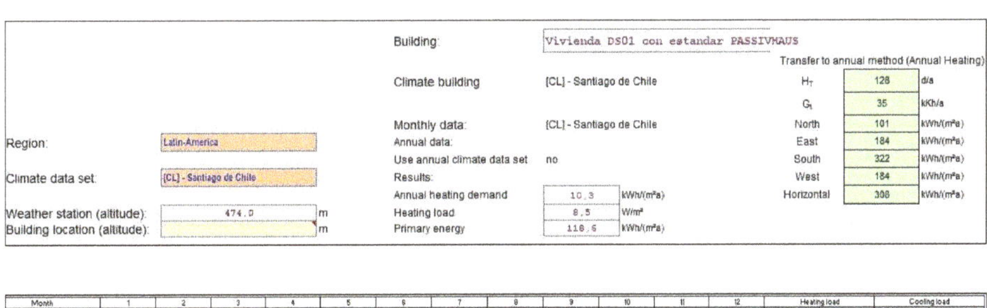
Building:	Vivienda DS01 con estandar PASSIVHAUS		
Climate building	[CL] - Santiago de Chile		
Monthly data:	[CL] - Santiago de Chile		
Annual data:			
Use annual climate data set	no		
Results:			
Annual heating demand	10,3	kWh/(m²a)	
Heating load	8,5	W/m²	
Primary energy	118,6	kWh/(m²a)	

Region:	Latin-America	
Climate data set:	[CL] - Santiago de Chile	
Weather station (altitude):	474,0	m
Building location (altitude):		m

Transfer to annual method (Annual Heating)

H_T	128	d/a
G_t	35	kKh/a
North	101	kWh/(m²a)
East	184	kWh/(m²a)
South	322	kWh/(m²a)
West	184	kWh/(m²a)
Horizontal	308	kWh/(m²a)

Month	1	2	3	4	5	6	7	8	9	10	11	12	Heating load		Cooling load	
Days	31	28	31	30	31	30	31	31	30	31	30	31	Weather 1	Weather 2	Weather 1	Weather 2
[CL] - Santiago de Chile	Latitude:	33,4	Longitude:	-70,8	Altitude m	474	Daily temperature swing Summer (K)			15,7	Radiation data:	kWh/(m²month)	Radiation: W/m²		Radiation: W/m²	
Ambient temp	8,4	10,0	11,7	14,6	17,4	19,7	21,0	20,3	18,6	14,5	10,9	9,0	4,7	10,4	30,0	30,0
North	23	29	38	51	64	78	72	50	44	33	26	20	32	15	93	93
East	43	51	74	104	123	135	137	106	102	62	44	34	63	14	195	195
South	86	94	93	89	69	61	66	80	110	102	82	71	117	13	96	96
West	40	58	70	94	119	127	125	108	98	68	43	35	54	14	181	181
Global	68	91	124	176	219	246	243	194	169	109	74	56	92	27	349	349
Dew point	4,5	5,8	6,4	7,8	7,9	8,7	9,2	9,7	9,6	7,7	6,2	5,4			12,7	12,7
Sky temp	-2,3	-0,9	0,1	1,9	1,4	3,5	4,1	5,2	3,6	2,1	0,6	-0,8			8,8	12,7
Ground temp	16,0	15,7	15,8	16,3	17,1	20,0	20,7	21,0	20,9	18,4	17,6	16,7	15,7	15,7	21,0	21,0
Comment:		Location in the south hemisphere. Data set adaptad for use in PHPP (see PHPP user's manual)!														

Figure A2. Climate.

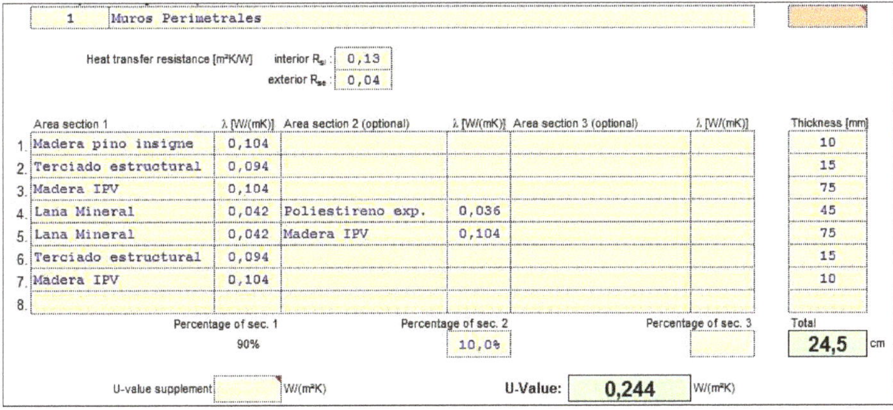
1 Muros Perimetrales

Heat transfer resistance [m²K/W] interior R_{si}: 0,13 exterior R_{se}: 0,04

	Area section 1	λ [W/(mK)]	Area section 2 (optional)	λ [W/(mK)]	Area section 3 (optional)	λ [W/(mK)]	Thickness [mm]
1.	Madera pino insigne	0,104					10
2.	Terciado estructural	0,094					15
3.	Madera IPV	0,104					75
4.	Lana Mineral	0,042	Poliestireno exp.	0,036			45
5.	Lana Mineral	0,042	Madera IPV	0,104			75
6.	Terciado estructural	0,094					15
7.	Madera IPV	0,104					10
8.							
	Percentage of sec. 1	90%	Percentage of sec. 2	10,0%	Percentage of sec. 3		Total **24,5** cm

U-value supplement W/(m²K) **U-Value:** **0,244** W/(m²K)

Figure A3. *Cont.*

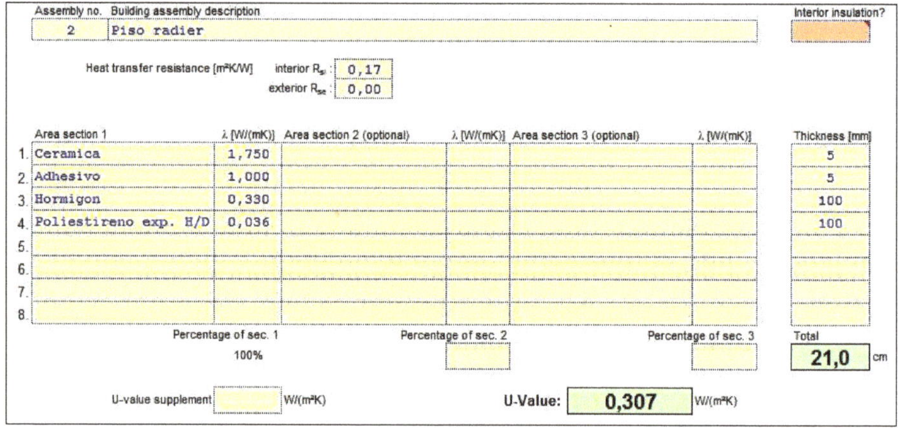

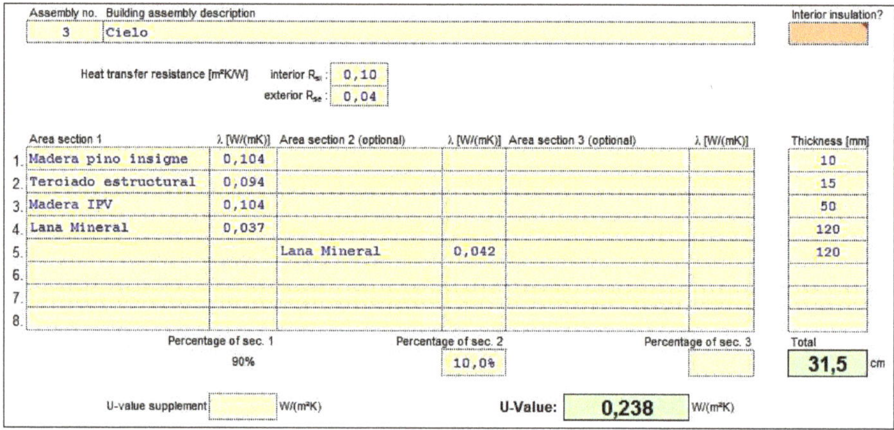

Figure A3. U Values.

Thermal bridge inputs													
Nr.	Thermal bridge description	Group Nr.	Assigned to group	Quantity	x (	User determined length [m]	-	Subtraction user-determined length [m]	)=	Length l [m]	Input of thermal bridge heat loss coefficient W/(m K)	Ψ W/(mK)
1	Corner of Wall	15	Thermal bridges Ambient	1	x (	9,70	-		) =	9,70	Corner of Wall	-0,164
2	Wall and roof	15	Thermal bridges Ambient	1	x (	43,00	-		) =	43,00	Wall and roof	-0,093
3	Encounter Floor- Wall	15	Thermal bridges Ambient	1	x (	43,00	-		) =	43,00	Encounter Floor- Wall	-0,003
4	Window frame	15	Thermal bridges Ambient	1	x (	33,80	-		) =	33,80	Window frame	-0,850
5	Door Frame	15	Thermal bridges Ambient	1	x (	9,60	-		) =	9,60	Door Frame	0,109
6	Door - floor junction	15	Thermal bridges Ambient	1	x (	1,60	-		) =	1,60	Door – floor junction	-0,159

Figure A4. Thermal Bridges.

Climate:	[CL] - Santiago de Chile										
Window area orientation	Global radiation (cardinal points)	Shading	Dirt	Non-perpendicular incident radiation	Glazing fraction	g-Value	Solar irradiation reduction factor	Window area	Window U-Value	Glazing area	Average global radiation
maximum:	kWh/(m²a)	0,75	0,95	0,85				m²	W/(m²K)	m²	kWh/(m²a)
North	101	0,52	0,95	0,85	0,490	0,77	0,21	1,45	2,03	0,71	101
East	184	0,66	0,95	0,85	0,552	0,77	0,29	2,40	2,14	1,32	184
South	322	0,77	0,95	0,85	0,699	0,77	0,44	9,00	2,25	6,30	322
West	184	0,66	0,95	0,85	0,490	0,77	0,26	1,45	2,03	0,71	184
Horizontal	308	1,00	0,95	0,85	0,000	0,00	0,00	0,00	0,00	0,00	308
		Total or average value for all windows.				0,77	0,37	14,30	2,18	9,04	

Figure A5. Windows.

Orientation	Glazing area	Reduction factor winter	Reduction factor summer
	m²	r_{Sw}	r_{Ss}
North	0,71	52%	62%
East	1,32	66%	72%
South	6,30	77%	60%
West	0,71	66%	72%
Horizontal	0,00	100%	100%

Figure A6. Shading.

Heating, cooling, DHW, auxiliary electricity, lighting, electrical appliances		65,6	118,6	29,7
Total PE Value	118,6	kWh/(m²a)		
Total emissions CO_2 Equivalent	29,7	kg/(m²a)		(Yes/No)
Primary Energy Requirement		120	kWh/(m²a)	yes

Heating, DHW, auxiliary electricity (no lighting and electrical appliances)		38,9	49,4	11,6
Specific PE demand - Mechanical system	49,4	kWh/(m²a)		
Total emissions CO_2-equivalent	11,6	kg/(m²a)		

Figure A7. Primary Energy.

References

1. Williams, J.; Mitchell, R.; Raicic, V.; Vellei, M.; Mustard, G.; Wismayer, A.; Yin, X.; Davey, S.; Shakil, M.; Yang, Y.; et al. Less is more: A review of low energy standards and the urgent need for an international universal zero energy standard. *J. Build. Eng.* **2016**, *6*, 65–74. [CrossRef]
2. Martinez-Soto, A.; Jentsch, M.F. Comparison of prediction models for determining energy demand in the residential sector of a country. *Energy Build.* **2016**, *128*, 38–55. [CrossRef]
3. División de Prospectiva y Política Energética del Ministerio de Energía (DPPE). Balance Nacional de Energía (1990–2013). Santiago, Chile. 2015. Available online: http://www.minenergia.cl/archivos_bajar/LIBRO-ENERGIA-2050-WEB.pdf (accessed on 25 July 2019).
4. Wegertseder, P.; Schmidt, D.; Hatt, T.; Saelzer, G.; Hempel, R. Barreras y oportunidades observadas en la incorporación de estándares de alta eficiencia energética en la vivienda. *Arquit. Urban.* **2014**, *15*, 37–49.
5. Corporación de Desarrollo Tecnológico (CDT). Estudio de Usos Finales y Curva de Oferta de la Conservación de la Energía en el Sector Residencial. Santiago, Chile. 2010. Available online: http://dataset.cne.cl/Energia_Abierta/Estudios/Minerg/Usos%20finales.pdf (accessed on 25 July 2019).
6. Hatt, T.; Saelzer, G.; Hempel, R.; Gerber, A. Alto confort interior con mínimo consumo energético a partir de la implementación del estándar Passivhaus en Chile. *Rev. Constr.* **2012**, *11*, 123–134. [CrossRef]
7. Simian, J.M. Logros y Desafíos de La Política Habitacional en Chile. *Estudios Públicos* **2010**, *1*, 117. Available online: https://www.cepchile.cl/cep/site/artic/20160304/asocfile/20160304095214/rev117_simian.pdf (accessed on 25 July 2019).
8. Martínez, P.; Sarmiento, P.; Urquieta, W. Evaluación de la humedad por condensación dentro de viviendas sociales. *Revista INVI* **2009**, *20*, 154–165. Available online: http://revistainvi.uchile.cl/index.php/INVI/article/view/323/887 (accessed on 25 July 2019).
9. Subhashini, S.; Thirumaran, K. A passive design solution to enhance thermal comfort in an educational building in the warm humid climatic zone of Madurai. *J. Build. Eng.* **2018**, *18*, 395–407. [CrossRef]
10. Golshan, M.; Thoen, H.; Zeiler, W. Dutch sustainable schools towards energy positive. *J. Build. Eng.* **2018**, *19*, 161–171. [CrossRef]
11. Košir, M.; Gostiša, T.; Kristl, Ž. Influence of architectural building envelope characteristics on energy performance in Central European climatic conditions. *J. Build. Eng.* **2017**, *15*, 278–288. [CrossRef]
12. Scoccia, R.; Toppi, T.; Aprile, M.; Motta, M. Absorption and compression heat pump systems for space heating and DHW in European buildings: Energy, environmental and economic analysis. *J. Build. Eng.* **2018**, *16*, 94–105. [CrossRef]

13. Venkatesan, K.; Ramachandraiah, U. Climate responsive cooling control using artificial neural networks. *J. Build. Eng.* **2018**, *19*, 191–204. [CrossRef]
14. Short, C.A. The recovery of natural environments in architecture: Delivering the recovery. *J. Build. Eng.* **2018**, *15*, 328–333. [CrossRef]
15. Eyre, N.; Baruah, P. Uncertainties in future energy demand in UK residential heating. *Energy Policy* **2015**, *87*, 641–653. [CrossRef]
16. McKenna, R.; Merkel, E.; Fehrenbach, D.; Mehne, S.; Fichtner, W. Energy efficiency in the German residential sector: A bottom-up building-stock-model-based analysis in the context of energy-political targets. *Build. Environ.* **2013**, *62*, 77–88. [CrossRef]
17. Nik, V.M.; Mata, É.; Kalagasidis, A. A statistical method for assessing retrofitting measures of buildings and ranking their robustness against climate change. *Energy Build.* **2015**, *88*, 262–275. [CrossRef]
18. Ministerio de Vivienda y Urbanismo (MINVU). Ordenanza General de Urbanismo y Construciones, Santiago, Chile. 2018. Available online: https://www.minvu.cl/elementos-tecnicos/normativa-de-urbanismo-y-construcciones/d-s-n47-de-1992-ordenanza-general-de-urbanismo-y-construcciones/ (accessed on 25 July 2019).
19. Passivhaus Institut (PHI). About Passive House—What Is a Passive House? Darmstadt, Germany. 2018. Available online: https://passipedia.org/basics/what_is_a_passive_house (accessed on 25 July 2019).
20. Pitts, A. Passive house and low energy buildings: Barriers and opportunities for future development within UK practice. *Sustainability* **2017**, *9*, 272. [CrossRef]
21. Minergie. Baustandard Minergie-P. 2018. Available online: https://www.minergie.ch/de/verstehen/baustandards/ (accessed on 25 July 2019).
22. Schnieders, J.; Feist, W.; Rongen, L. Passive Houses for different climate zones. *Energy Build.* **2015**, *105*, 71–87. [CrossRef]
23. Tatarestaghi, F.; Ismail, M.A.; Ishak, N.H. A comparative study of passive design features/elements in Malaysia and passive house criteria in the tropics. *J. Des. Built Environ.* **2018**, *18*, 15–25. Available online: https://ejournal.um.edu.my/index.php/jdbe/article/view/15577 (accessed on 14 July 2019).
24. Truong, H.; Garvie, A.M. Chifley Passive House: A Case Study in Energy Efficiency and Comfort. *Energy Procedia* **2017**, *121*, 214–221. [CrossRef]
25. Badescu, V.; Rotar, N.; Udrea, I. Considerations concerning the feasibility of the German Passivhaus concept in Southern Hemisphere. *Energy Effic.* **2015**, *8*, 919. [CrossRef]
26. Tataru, A. Possibility to Implement the Concept of Passivhaus in Russia. In Proceedings of the 18th International Multidisciplinary Science GeoConference: Ecology, Economics, Education and Legislation (SGEM 2018), Sofia, Bulgaria, 30 June–9 July 2018; pp. 723–731.
27. Liang, X.; Wang, Y.; Royapoor, M.; Wu, Q.; Roskilly, T. Comparison of building performance between Conventional House and Passive House in the UK. *Energy Procedia* **2017**, *142*, 1823–1828. [CrossRef]
28. Elsarrag, E.; Alhorr, Y. Modelling the thermal energy demand of a Passive-House in the Gulf Region: The impact of thermal insulation. *Int. J. Sustain. Built Environ.* **2012**, *1*, 1–15. [CrossRef]
29. Mihai, M.; Tanasiev, V.; Dinca, C.; Badea, A.; Vidu, R. Passive house analysis in terms of energy performance. *Energy Build.* **2017**, *144*, 74–86. [CrossRef]
30. Udrea, I.; Croitoru, C.; Nastase, I.; Crutescu, R.; Badescu, V. Thermal Comfort in a Romanian Passive House: Preliminary Results. *Energy Procedia* **2016**, *85*, 575–583. [CrossRef]
31. Galvin, R. Are passive houses economically viable? A reality-based, subjectivist approach to cost-benefit analyses. *Energy Build.* **2014**, *80*, 149–157. [CrossRef]
32. Dalbem, R.; Cunha, E.G.; Vicente, R.; Figueiredo, A.; Oliveira, R.; Da Silva, A.C.S.B. Optimisation of a social housing for south of Brazil: From basic performance standard to passive house concept. *Energy* **2019**, *167*, 1278–1296. [CrossRef]
33. Chovancová, J.; Kocourková, G.; Kozumplíková, L. Assessment of operating cash flow of the investment in a construction of passive houses. *Tehnički Glasnik* **2014**, *8*, 339–346.
34. Zhang, X.; Platten, A.; Shen, L. Green property development practice in China: Costs and barriers. *Build. Environ.* **2011**, *46*, 2153–2160. [CrossRef]
35. Coyle, D. An investigation into the cost optimality of the Passive House retrofi t standard for Irish dwellings using Life Cycle Cost Analysis. *J. Sustain. Des. Appl. Res.* **2016**, *4*, 4–14. [CrossRef]
36. Shim, J.; Song, D.; Kim, J. The economic feasibility of passive houses in Korea. *Sustainability* **2018**, *10*, 3558. [CrossRef]

37. MINVU. *Estadísticas de Subsidios Otorgados Programa Regular y Reconstrucción, Equipo de estadísticas- Comisión de Estudios Habitacionales y Urbanos: Periodo 1990–Septiembre 2017*; Ministerio de Vivienda y Urbanismo: Santiago, Chile, 2017. Available online: https://www.minvu.cl/elementos-tecnicos/estadisticas/viviendas-contratadas-y-subsidios-otorgados/ (accessed on 25 July 2019).
38. DGA. *Anuario Climatológico*; Dirección General Aeronáutica: Santiago, Chile, 2010. Available online: https://climatologia.meteochile.gob.cl/ (accessed on 25 July 2019).
39. Ministerio de Energia (MINERGIA). Explorar Recurso Solar y Datos Meteorológicos. Santiago, Chile. 2019. Available online: http://www.minenergia.cl/exploradorsolar/ (accessed on 7 July 2019).
40. Banco Central de Chile. Índice de Precios de Vivienda en Chile: Metodología y Resultados. *Estudios Económicos Estadísticos* **2014**, *107*, 1. Available online: https://www.bcentral.cl/web/guest/-/indice-de-precios-de-vivienda-en-chile-metodologia-y-resultad-1 (accessed on 25 July 2019).
41. Instituto Nacional Estadísticas (INE). Síntesis de Resultados ESI, Encuesta Suplementaria de Ingresos 2017. Santiago, Chile. 2018. Available online: https://www.ine.cl/docs/default-source/ingresos-y-gastos/esi/ingreso-de-hogares-y-personas/resultados/2017/sintesis_esi_2017_nacional.pdf (accessed on 25 July 2019).
42. Construcción Chile. Precios de Construcción de Casas y Galpones. Santiago, Chile. 2019. Available online: https://www.construccionchile.cl/contacto (accessed on 5 August 2020).
43. Ministerio del Mediambiente, MMA. Plan de Descontaminación Atmosférica MP 2,5. Decreto Supremo número 8. Santiago, Chile. 2015. Available online: https://www.bcn.cl/leychile/navegar?idNorma=1084085 (accessed on 5 August 2020).
44. Instituto Nacional de Estadisticas, INE. Medio Ambiente Informe Anual 2014. Santiago, Chile. 2015. Available online: https://www.ine.cl/estadisticas/economia/energia-y-medioambiente/variables-basicas-ambientales (accessed on 5 August 2020).
45. Servicios Climáticos. Dirección meteorológica de Chile. Santiago, Chile. 2018. Available online: https://climatologia.meteochile.gob.cl/application/index/anuarios (accessed on 5 August 2020).
46. Hatt, T. El estándar 'passivhaus' en el centro-sur de chile un estudio parámetrico multifactorial. Ph.D. Thesis, University of Bio-Bio, Concepción, Chile, 2012.
47. Avendaño-Vera, C.; Martinez-Soto, A.; Marincioni, V. Determination of optimal thermal inertia of building materials for housing in different Chilean climate zones. *Renew. Sustain. Energy Rev.* **2020**, 131. [CrossRef]
48. Feist, W. *Passivhaus Definition*; Passive House Institute: Darmstadt, Germany, 2005. Available online: https://passiv.de/former_conferences/neunte/Passivhaus_Definition.html (accessed on 25 July 2019).
49. ONDAC. Manual de Precios de la Construcción. Santiago, Chile. 2019. Available online: www.ondac.com (accessed on 25 July 2019).

Publisher's Note: MDPI stays neutral with regard to jurisdictional claims in published maps and institutional affiliations.

Article

Air Changes for Healthy Indoor Ambiences under Pandemic Conditions and Its Energetic Implications: A Galician Case Study

José A. Orosa [1,*], Modeste Kameni Nematchoua [2,3] and Sigrid Reiter [3]

1 Department of Nautical Science and Marine Engineering, E.T.S.N. y M., University of A Coruña, Paseo de Ronda 51, 15011 A Coruña, Spain
2 Beneficiary of an AXA Research Fund Postdoctoral Grant, Research Leaders Fellowships, AXA SA 25 Avenue Matignon, 75008 Paris, France; mkameni@uliege.be
3 LEMA, ArGEnCo Department, University of Liège, 4000 Liège, Belgium; sigrid.reiter@uliege.be
* Correspondence: jaorosa@udc.es

Received: 16 September 2020; Accepted: 12 October 2020; Published: 14 October 2020

Featured Application: An analysis of the optimal ventilation level in naturally ventilated classrooms under COVID-19 pandemic requirements.

Abstract: The present paper aims to show a mathematical understanding of the effect of ventilation rate over building energy consumption. Moreover, as a case study to show this methodology, a proposal was analyzed of modifying the teaching period to reach a maximum increase of air changes in school buildings, to allow adherence to the COVID-19 pandemic requirements in the Galicia region, with lower energy consumption. In this sense, to analyze the energetic implication of this proposal, the building construction was defined, modeled in accordance with the ISO Standard 13790 and implemented in accordance with the Monte Carlo method. Results showed the probability of energy consumption as a Weibull model. Furthermore, a map of different Weibull models in accordance with different ventilation rates was developed. The constants of the Weibull models allow to identify normal distributions of the probability density functions of energy consumption, especially the ones with lower energy consumption. As a consequence, these constants are a better parameter to identify the optimal ventilation rate for each season in search of a healthy indoor ambience, which is of interest for a future design guide. Finally, the main results showed a reduction of energy consumption at a higher ventilation rate in the summer season. As a consequence, the necessity of modifying teachings periods, as an adequate procedure to prevent more COVID infections, is concluded.

Keywords: ventilation; energy; COVID-19; procedure; building; ISO

---

1. Introduction

Indoor ambiences are controlled to reach an adequate thermal comfort, ensuring healthy conditions. In this sense, in mild climates with high relative humidity, such as in the Galicia region, there is no need of complex Heating Ventilation and Air Conditioning Systems (HVAC) in most public buildings, such as schools and offices. Most of the time, solely a hot water system fed by a fuel oil boiler is employed during the coldest winter days, with an average relative humidity of 80%. Moreover, the ventilation rate is manually controlled by opening windows in accordance with common sense and the occupant's perception of indoor air quality. This perception, in accordance with previous research works [1], depends on the time that an occupant stays in that indoor environment, which reaches the lowest perception of air quality after some minutes in that environment. As a consequence, the need for natural ventilation and air renovation arises, most of the times in public buildings, when the level of indoor quality has worsened significantly.

ASHRAE [2] has proposed increasing the ventilation rate due to the COVID-19 pandemic. In this sense, "transmission of COVID-19 through the air is sufficiently likely that airborne exposure to the virus should be controlled. Changes to building operations, including the operation of heating, ventilating, and air-conditioning (HVAC) systems, can reduce airborne exposures" [2]. All these changes must be done in agreement with the ASHRAE Standard 62.1-2019. In particular, some designer guidelines for general schools proposed by the ASHRAE recommend that relative humidity in the winter season must not decrease below 50% due to that it increases the probability of COVID-19 propagation, which is not a problem for the humid region of Galicia. Moreover [3], it recommends installing CO_2 sensors at the occupied zone that warn against underventilation and, during the epidemic, to change the default limits to 800 ppm (warning) and 1000 ppm (alarm). Moreover, if mechanical ventilation is not employed, it is proposed to employ HEPA air purifiers. Despite this, an adequate air renovation and dilution of contaminants are required. As a consequence, it is recommended to ensure that the airflow patterns in classrooms are adjusted to minimize occupant exposure to particles [2]. As a consequence, different laminar air movements are required to prevent aerosols, which create a high to spread the COVID-19 through air flow [4]. Moreover, all these comments are in agreement with the World Health Organization (WHO) [5] proposing to increase ventilation when the climate allows.

Finally, energy consumption is one of the most important factors to be considered during the designing of new constructions, and these need to be improved in the existing buildings, as was shown in recent research works [6–11]. In this sense, there are a lot of software resources that can be used to estimate building energy consumption during the full year and different standards for building energetic qualification [12,13]. One of the main objectives of the International Energy Agency [14] is to obtain new methodologies for defining better alternatives for preventing energy consumption in buildings; there is currently a knowledge gap in this field. Currently, there is no real understanding of the ventilation in buildings, which reduces the options of improvement proposals. As a consequence, the aim of this work is to analyze the energetic effect of the air changes in buildings and to identify the guide parameters to increase the ventilation rate. In particular, a case study of this procedure in a Galician school building, in accordance with the pandemic requirements, was shown. In this sense, a calculation procedure based on ISO Standard 13790 was developed in accordance with the Monte Carlo method [15–17] with the aim to obtain an original result not shown by software resources of building energy rating; additionally, a probabilistic model for energy use was employed in accordance with a Weibull distribution [18–22]. Furthermore, this probabilistic model will be employed to define the expected energy consumption at different ventilation rates and weather conditions and to identify the better options to reach a healthy indoor ambience with lower energy consumption.

2. Materials

2.1. Climatic Data

Climatic data were collected through a sampling process in 50 weather stations located in central and remarkable points in Galicia. These weather stations sample temperature, relative humidity, and wind speed variables with a sampling frequency from about 5 to 10 min [12,23] and an error margin of 0.1 °C, 0.2%, and 0.1 m/s, which can be considered acceptable. Moreover, these weather stations were chosen for the purpose of this study because they ignore buildings and other conditions that could possibly interfere with the data.

2.2. School Buildings

The school building selected for this case study is placed in Galicia, in the northwest of Spain. In this school building, the average classroom occupation level is 25 occupants, with an average floor area of 50 m^2, in accordance with the COVID-19 pandemic requirements based on the Galician Ministry recommendations.

The school was built using concrete with walls of two brick layers thick, and an internal air barrier. The floor is in direct contact with the ground, and not insulated. Concurring with earlier research works [24], the internal heat capacity of the building was calculated in accordance with the EN ISO 13790 standard, and the resulting value was found to be 150 kJ/(m^2 K). This value corresponds to a medium heavy structure. Finally, the U-value for the windows is 1.8 W/m^2K, in accordance with the low emissive and double-glazed windows.

In line with previous research works on thermal inertia, typical wall constructions were defined, and minimum and maximum wall thermal inertia were defined in accordance with the standards. In light of this, values between 150 kJ/(m^2 K) and 400 kJ/(m^2 K) were selected to be employed in the next simulation processes.

Finally, this classroom has natural ventilation through windows that can be opened manually, reaching an average ventilation level of 0.8 air changes per hour in winter and summer seasons. This value was obtained with the tracer gas decay method, employing SF6 as sampled gas. The geometry of the typical school building is shown in Figure 1, and shows two zones: the professor's zone and the student's zone.

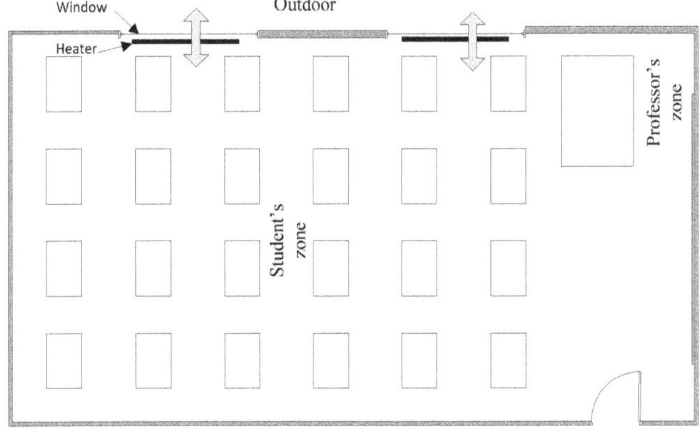

Figure 1. Typical school building plant.

2.3. Internal Heat Gains

The proposed values of internal heat gains in the EN ISO 13790 standard can be used if no national data are available. Therefore, the simulations were adapted to the actual situation. The sum of the internal heat gains from occupants, lighting, and devices gives the annual internal heat gains, which are proposed in the standard, as constant values 5.7 W/m^2 for school buildings. In our case study, concurring with previous research works [25], the real annual internal heat gain was analyzed and fixed at 44 kWh/m^2 year.

2.4. HVAC System Set Point Temperature

The heating system typically employed in Galician classrooms consist of a central heating system consisting of a fuel boiler that heats water at 80 °C and, once mixed with return water, is circulated in different classrooms. In particular, the building objective of this study is heated continuously for about 4 h, which represents part of the working period. In this sense, in accordance with the ISO Standard recommendations for this region, the set point of heating and cooling is 18 °C and 23 °C, respectively, to maintain better thermal comfort while maintaining low energy consumption.

3. Methods

As was commented before, the methodology employed in the present study aims to define the probability density function of energy consumption during the winter and summer seasons in buildings. With this information, different ventilation levels of the classroom in the Galician region will be analyzed to identify the real effects and requirements of a COVID-19 pandemic situation.

Despite the fact that all the methods are explained in this section, an in-depth definition of the Monte Carlo method over ISO 13790, taking as reference some residential building, can be seen in previous work [26]. Moreover, in this previous research, the results obtained with this methodology were validated with respect to the certified energy rating software resources.

3.1. ISO 13790 Standard

This ISO 13790 Standard was developed by the European Committee for Standardization (CEN), in collaboration with the International Organization for Standardization (ISO) to define the calculation methodology of the energy consumption and carbon dioxide emissions in buildings. Despite the fact that ISO employs different levels of calculation complexity, for the present paper, the annual energy use method was employed.

The process begins with a selection of the main variables that must be considered in the building's energy consumption. In this sense, variables such as indoor air temperature in winter and summer seasons, building thermal inertia, air changes and the number of occupants were selected in accordance with ISO Standard 13790 and adjusted to the values of a real building [27–33]. From any modification of these values, when the outdoor air temperature is entered in a random way in accordance with the Monte Carlo Method procedure, the probability distribution of the expected energy consumption due to each building modification is obtained. In the base case, the number of air changes was 0.8, thermal inertia was 150 kJ/m^2 K, and the number of occupants was 25. At the same time, the average outdoor temperature conditions were defined in accordance with the ISO Standard indications for the region objective of this study. The heating and cooling systems, with set point temperatures of 18 °C and 23 °C, were identified with energy performance indices of 0.7 and 2.55, respectively. Subsequently, once the random input data were entered, the corresponding histograms were developed, as shown in Figure 2, and the probability density function of heating and cooling energy consumption for different air change values was defined, as shown in Figures 3–8. Finally, it is of interest to highlight that, as shown in previous works, the effect of thermal inertia does not exist in these kinds of yearly studies about energy consumption.

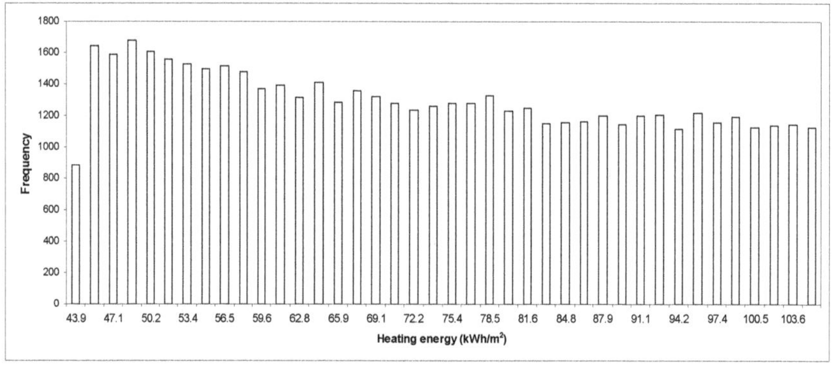

Figure 2. Histogram of energy consumption during the heating season.

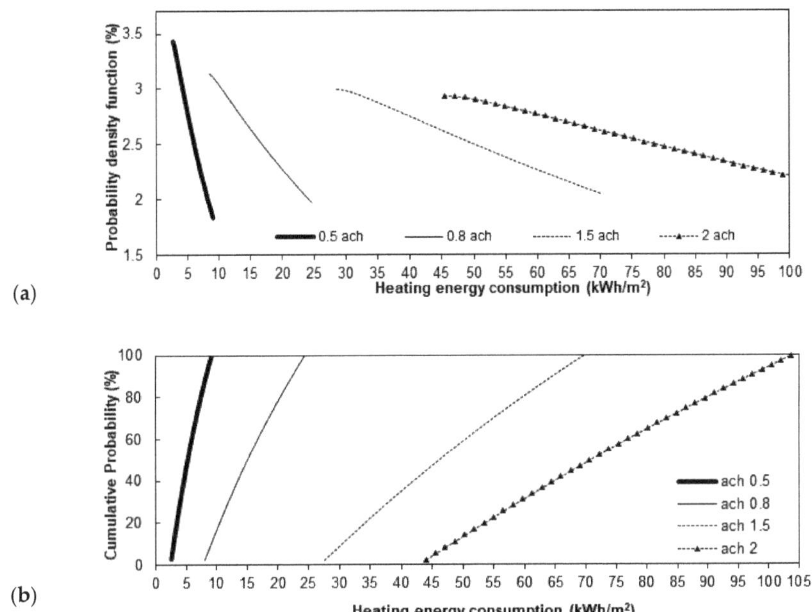

Figure 3. Probability density functions (**a**) and cumulative probability (**b**) of the heating energy consumption for different air changes (Ach = 0.5, Ach = 0.8, Ach = 1.0, Ach = 2.0).

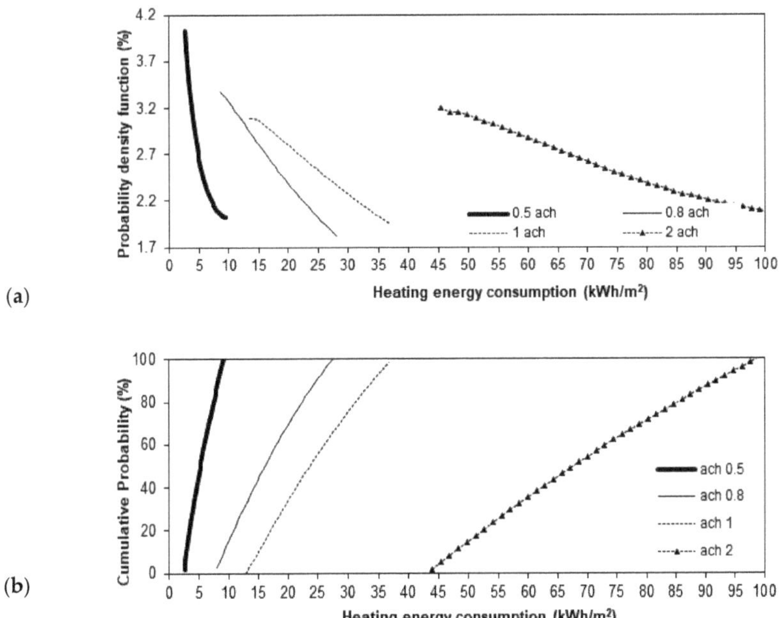

Figure 4. Probability density functions (**a**) and cumulative probability (**b**) of the heating energy consumption for an increment of +0.5 °C in the outdoor temperature.

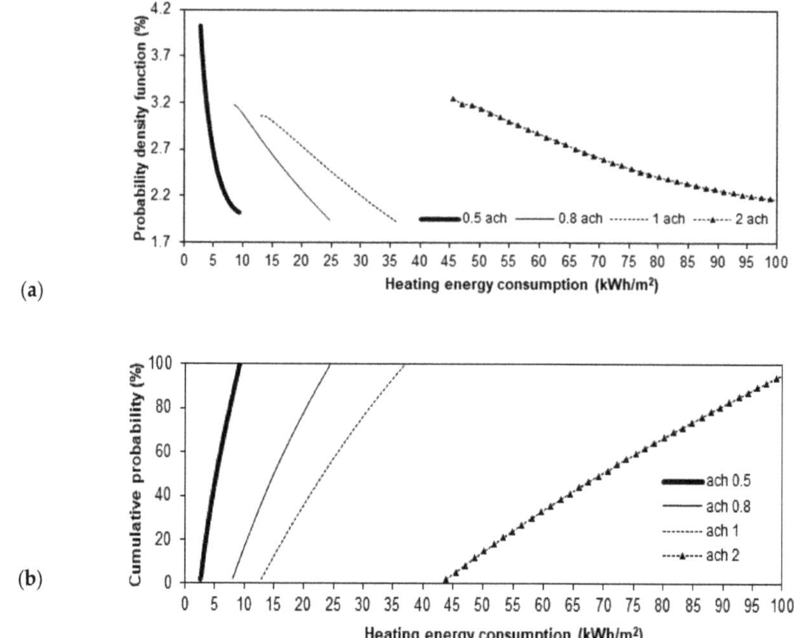

Figure 5. Probability density functions (**a**) and cumulative probability (**b**) of the heating energy consumption for an increment of +1 °C in the outdoor temperature.

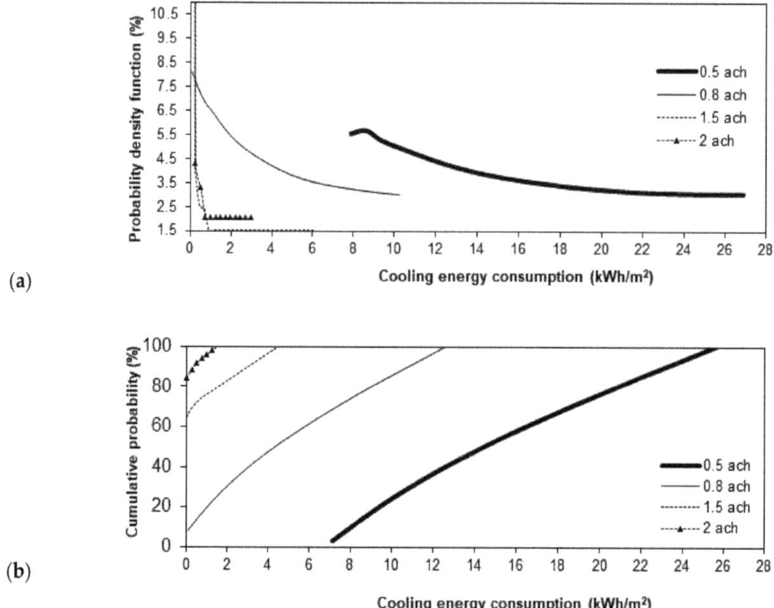

Figure 6. Probability density functions (**a**) and cumulative probability (**b**) of the cooling energy consumption for different air changes (Ach = 0.5, Ach = 0.8, Ach = 1.0, Ach = 2.0).

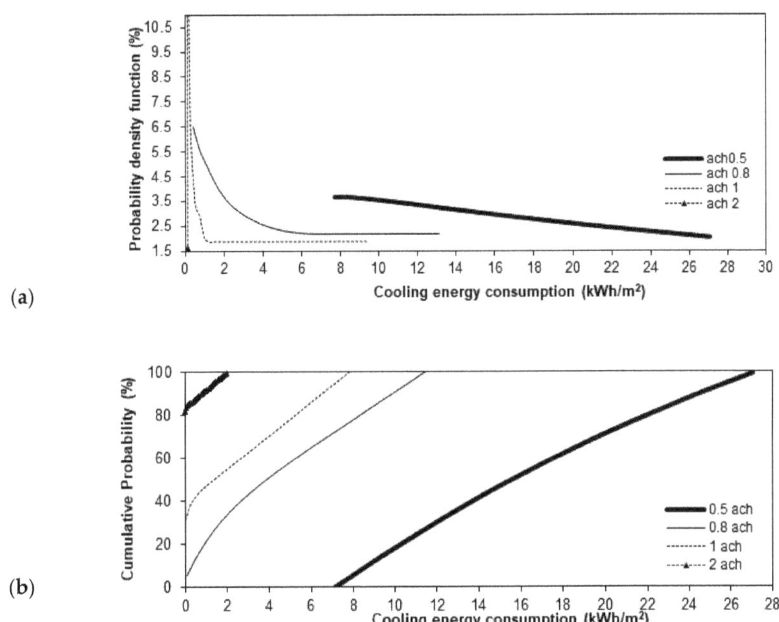

Figure 7. Probability density functions (**a**) and cumulative probability (**b**) of the cooling energy consumption for an increment of +0.5 °C in the outdoor temperature.

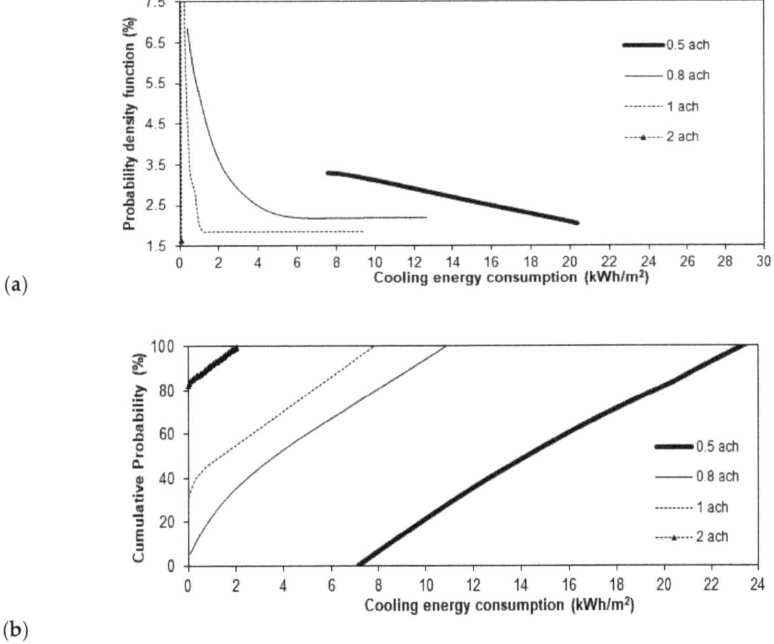

Figure 8. Probability density functions (**a**) and cumulative probability (**b**) of the cooling energy consumption for an increment of +1 °C in the outdoor temperature.

3.2. Calculation Process: Number of Iterations

The random process was developed in Visual Basic for Applications for each different curve to obtain an adequate determination factor during the curve fitting process. In this sense, the Monte Carlo method predicts the estimation of the total error (ε), which is proportional to the number of iterations, by Equation (1):

$$\varepsilon = \frac{3\sigma}{\sqrt{N}} \tag{1}$$

where σ is the standard deviation of the random variable and N is the number of iterations.

For instance, in the present case study, the standard deviation of the outdoor temperature was 1.45 °C in the summer and 2.09 °C in the winter season and the minimum number of iterations required for an error less than 2% was calculated. As a consequence, the minimum number of iterations proposed for each season was 166 in winter and 86 in summer season. Due to the fact that the ISO procedure needs few calculations and the curve fitting needs the highest number of values possible, an iterative process was developed with more than 60,000 iterations.

However, after a curve fitting in accordance with more than 60,000 mathematical models, a Weibull density distribution was selected in accordance with the determination factor obtained. The curve fitting process was done with the CFtool of Matlab software. Despite this, the codification was implemented by the authors to develop this task automatically, and to order the obtained curves by its determination factor. Within the 10 more accurate equations obtained, the Weibull model was selected due to its well-known relation with processes that change its probability density function during its life, as it used to be in maintenance models.

As a consequence, this adequate curve fitting could be identified by a determination factor of 0.90 between the simulated data and the values defined by the Weibull model of Equation (2):

$$y = a_0 + b_0 \cdot e_0^{\left(-\left(\left(\frac{x-c_0}{d_0}\right)+\left(\frac{e_0-1}{e_0}\right)^{\frac{1}{e_0}}\right)^{e_0} + \left(\frac{e_0-1}{e_0}\right)\right)} \cdot \left(\frac{e_0-1}{e_0}\right)^{-\left(\frac{e_0-1}{e_0}\right) \cdot \left(\left(\frac{x-c_0}{d_0}\right)+\left(\frac{e_0-1}{e_0}\right)^{\frac{1}{e_0}}\right)^{e_0-1}} \tag{2}$$

where x is the input variable and a_0, b_0, c_0, d_0, and e_0 are the model constants.

4. Results and Discussion

Once the Monte Carlo Method was applied in the calculation procedure defined by the building Standard ISO 13790, with more than 60,000 runs, the frequency of each different value was defined, as shown in Figure 2.

This histogram could be curve fitted for each different study in accordance with the Weibull distribution model of Equation (2) with an adequate determination factor, as can be observed in Figure 3.

After simulating the actual conditions, based on the values obtained from weather stations in Galicia and ISO 13790 standards indications, and in accordance with the yearly variation of the average temperature in winter and summer season of 0.5 °C and 1 °C [34], the Monte Carlo method was applied again with this temperature increase in the mean outdoor temperature. As a consequence, Figures 4–8 were obtained.

As an initial comparative parameter between Weibull models, we can use the value of the model constants shown in Tables 1 and 2 obtained from a curve fitting process for heating and cooling seasons at different average air temperatures and air changes. At the same time, the point of maximum probability value of each cumulative curve is of real interest, due to that it is the highest energy consumption possible; moreover, the maximum of the probability density function has been analyzed, as it represents the more frequent value of energy consumption.

Table 1. Curve fitting of probability density function during the heating season.

Average Heating Energy Consumption (kWh/m²)	ACH (h^{-1})	Outdoor Temp. (°C)	a_0	b_0	c_0	d_0	e_0	r^2
5.44	0.5	Average	3.363271	2.661263	9.437849	1.011474	0	0.92
5.10	0.5	+0.5 °C	1.854899	2.080979	2.565297	2.295291	1.007102	0.97
4.80	0.5	+1 °C	1.854899	2.080979	2.565297	2.295291	1.007102	0.97
15.72	0.8	Average	3.090611	8.386542	32.028872	1.010309	0	0.93
15.68	0.8	+0.5 °C	3.304539	8.415014	29.33913	1.012122	0	0.90
15.65	0.8	+1 °C	3.124542	8.369421	30.40935	1.010302	0	0.90
46.97	1.0	Average	3.212779	13.38395	42.19215	1.010308	0	0.91
24.20	1.0	+0.5 °C	3.077777	13.48641	45.60571	1.011666	0	0.92
22.19	1.0	+1 °C	3.050013	13.48202	45.16667	1.011673	0	0.92
72.96	2.0	Average	2.882833	46.047900	162.5114	1.012475	0	0.85
70.25	2.0	+0.5 °C	1.512973	1.519573	48.13923	39.73227	1.093759	0.91
68.92	2.0	+1 °C	1.862788	1.223835	47.25478	30.64512	1.095933	0.94

Table 2. Curve fitting of probability density function during the cooling season.

Average Cooling Energy Consumption (kWh/m²)	ACH (h^{-1})	Outdoor Temp. (°C)	a_1	b_1	c_1	d_1	e_1	r^2
15.97	0.5	Average	2.794215	2.479324	8.207154	5.324309	1.06168	0.91
16.34	0.5	+0.5 °C	3.618063	7.566651	29.62327	1.013869	0	0.83
16.94	0.5	+1 °C	3.273713	7.46334	24.62232	1.0127	0	0.84
5.30	0.8	Average	3.668553	9.12329	−2.019566	3.309106	1.0585379	0.95
5.45	0.8	+0.5 °C	3.273713	7.46334	24.62232	1.0127	0	0.84
5.52	0.8	+1 °C	1.973009	4.136657	0.203551	1.313985	1.085427	0.81
1.18	1.0	Average	2.734177	3321.353	−3.61981	3.85073	3.236254	0.99
3.16	1.0	+0.5 °C	3.077777	13.48641	45.60571	1.011666	0	0.92
4.38	1.0	+1 °C	3.050013	13.48202	45.16667	1.011673	0	0.92
0.36	2.0	Average	1.449848	347.4151	−0.27093	0.463775	3.049235	0.99
1.15	2.0	+0.5 °C	1.512973	1.519573	48.13923	39.73227	1.093759	0.91
1.19	2.0	+1 °C	0.447502	149105.2	−0.3577	1.293282	10.01984	0.99

In this sense, in Figure 3a the curve of the probability density function of the heating energy consumption after a 0.5 ach shows a more probable energy consumption of 2.89 kWh/m² with a frequency of 3.39% of the cases. Despite this, the energy consumption may reach 9.03 kWh/m², as we can see at the end of this same curve. If we now employ the cumulative curve of Figure 3b and we locate the cut of the 0.5 ach curve with the 100% cumulative probability, it can be concluded that all the energy consumption after a 0.5 ach will be equal to or lower than 9.03 kWh/m².

Despite the fact that these are really interesting parameters, a representative value is needed. As a consequence, numerical integration of the probability density functions led us to define the average energy consumption for each case. If we analyze the effect of the augmentation of 0.5 °C of the outdoor air temperature over this initial situation, see Figures 5 and 6 and Table 1, low modification can be observed with respect to the base case of the heating energy consumption under 0.5 ach. In particular, from these figures it can be observed that the highest energy consumption for the lowest air changes of 0.5 °C is about 9.03 kWh/m², like in previous cases. Despite this, its probability is reduced as outdoor temperature increases. For example, it is about 4.02% when the outdoor air increase from 0.5 °C to 1 °C. Thus, it can be concluded that this effect is being diffused when the number of air changes increases.

When we analyze the effect of air changes over the heating energy consumption, we can conclude from Figure 4a that when the number of air changes increases, the heating energy consumption increases. For example, in Figure 4a, we can see that the most probable energy consumption for 0.5 air changes is about 2.72 kWh/m^2 for a probability of 3.65%. At the same time, for two air changes, we have a probability of 3.09% for a heating energy consumption of 14.15 kWh/m^2.

Despite the fact that the summer season is a holiday, cooling systems may be added to classrooms due to the expected need to reduce pandemic infections in this hot period. As a consequence, the cooling energy demand estimated for that period was defined.

From Tables 1 and 2, it can be concluded that for the heating season, the energy consumption increases when the air changes increase. At the same time, the increment of outdoor temperature implies a reduction of the heating energy. In the cooling season, the energy consumption decreases at the same time that air changes increase. Moreover, it can be observed that energy consumption will increase at the same time that outdoor temperature increases. Despite this, the cooling energy consumption is, overall, reduced for that climatic region. Therefore, it might be interesting to consider changing part of the winter teaching period to the summer season due to the high energy saving and reduced possibility of contagion.

As the aim of this work is to define the optimal moment and the exact increment of the number of air changes to lower energy consumption, this will be solved numerically and by the proposed Weibull model constants. To solve it numerically, two models were obtained by curve fitting from the information shown in Tables 1 and 2. The first model, Equation (3), is for the heating period (HE) and was obtained with a determination factor of 0.95:

$$\mathrm{HE} = 15.7 - 24.9 \cdot \mathrm{ACH} - 9.6 \cdot \Delta t + 27.8 \cdot \mathrm{ACH}^2 + 11.1 \cdot \Delta t^2 - 7.38 \cdot \mathrm{ACH} \cdot \Delta t \quad (3)$$

where ACH refers to the air changes (h^{-1}) and Δt is the increase in temperature with regards to the seasonal average value (°C).

When the minimum of this function was solved numerically, a minimum energy consumption of 6.20 kWh/m^2 was obtained during the winter when the air changes are 0.52 h^{-1} and there is an increment of 0.60 °C with respect to the average seasonal value. The maximum of this function, with 77 kWh/m^2, was obtained at the average winter temperature and the number of air changes is 2 ach. This maximum energy consumption corresponds with the proposed option to prevent COVID infections during the winter season. As a consequence, since the objective is to increase the number of air changes, the same analysis was done in the summer season to define the cooling energy (CE) model, defined by Equation (4). This model was obtained with a determination factor of 88.15:

$$\mathrm{CE} = 29.35 - 35.41 \cdot \mathrm{ACH} + 1.24 \cdot \Delta t + 10.56 \cdot \mathrm{ACH}^2 - 0.68 \cdot \Delta t^2 + 0.62 \cdot \mathrm{ACH} \cdot \Delta \quad (4)$$

where ACH refers to the air changes (h^{-1}) and Δt is the increase in temperature with regards to the seasonal average value (°C).

Based on this cooling model, the minimum energy consumption is identified with an increment of 0.15 °C of the seasonal average temperature and with 1.67 ach air changes. As a consequence, this is the better solution for a compatible increment of air changes in accordance with COVID requirements and a lower energy consumption.

The previous resolution is just a numerical resolution of the problem, but it does not let researchers understand the reason why this is the optimal ventilation period compared to alternative options. As a consequence, an analysis of the Weibull model constants was done.

From Tables 1 and 2, it can be observed that most of the determination factors present a high value, showing a good agreement between the Weibull model and the probability of energy consumption. In this sense, it can be observed that the constants "e_0" and "e_1" (shape parameter) can be related with each type of curve and, as a consequence, its corresponding energy consumption. In particular, when the "e" constant reaches a zero value, the curve is exponential and, when it is near to 3.5, a normal distribution is expected and the energy consumption is at a minimum. This is due to that the area below the curve is smaller in a normal distribution with respect to an exponential or a lognormal curve. Finally, if the "e" constant increases to a value over 3.5, the energy consumption tends to increase.

Another interesting conclusion that can be obtained from these tables is that a temperature increase of 0.5 °C or 1 °C in winter season makes the "e_0" constant grow towards a normal distribution and, as a consequence, the energy consumption is reduced. The inverse effect is observed in the summer season, when the temperature increase reduces the "e_1" value and, as a consequence, increases energy consumption.

Based on this analysis, the optimal ventilation rate is identified when the "e" constant is near 3 in two cases in the summer season. These two cases are at an average outdoor temperature condition and 2 and 1 ach, respectively. As a consequence, these two are the first and second optimal energetic solutions for this problem, since in these cases a normal distribution is expected, and the exact ventilation rate must be selected in accordance with the ventilation needs. It is interesting to highlight that this normal distribution is due to a high number of iterations with a reduced difference between outdoor and indoor conditions. As a consequence, this objective can be obtained with different building design parameters, and can thus function as a new way to guide HVAC operators and designers to produce more optimized buildings.

This result is in agreement with the previous numerical resolution and, as a consequence, it can be concluded that these constants are a good guide to define the conditions with a lower energy consumption in accordance with different requirements.

In our particular case study, it is of interest to the autonomous community of Galicia to calculate the exact energy saving if they do change the teaching periods. In particular, the increase of the heating energy consumption when the air changes increase from their actual 0.8 ach to 2 ach in the winter season. As a consequence, if we employ the more unfavorable increase of energy consumption in the winter season, identified as the highest augmentation in energy consumption defined by Table 1, 57.24 kWh/m^2 is obtained. This increase in energy consumption does not reflect a relevant change for any outdoor air temperature increase. If a Heating Ventilation and Air Conditioning system (HVAC) is employed for conditioning the classes in the summer period, an increment of the energy consumption, at the time to increase the air changes from 0.8 ach to 2 ach, of 0.36 kWh/m^2 was obtained. As a consequence, it can be concluded that a decrement of the energy consumption of 72.6 kWh/m^2 when part of the winter teaching months is replaced to summer teaching months.

Finally, due to that the summer season are mild in the Galicia region and that this calculation considers a HVAC system to cool indoor ambience, if this equipment depreciates because it is not really employed, the energy saving will increase up to 72.96 kWh/m^2. As a consequence, changing the teaching months to the summer season and increasing the air changes is an interesting method to reduce the pandemic virus. This will be improved by the fact that in the summer months, a reduced risk of contagion of COVID-19 is expected.

5. Conclusions

In the present paper, a new method to understand building energy consumption and selecting the optimal period to increase the air changes in accordance with different needs is shown. In particular, results showed that the Monte Carlo method applied over ISO 13790 standards are a useful tool to define the probability of building energy consumption. Furthermore, it was obtained that the probability of building energy consumption can be defined by a Weibull model. After analyzing these Weibull model constants, the shape factor "e" lets us identify the conditions at which the probability density function is a normal distribution. This normal distribution is one with a lower energy consumption and can be identified by an "e" of 3.5, which is a useful tool for buildings and HVAC designers.

In particular, this methodology was employed to analyze the case study of energy saving when some teaching periods are changed from winter to summer seasons. It was obtained that, due to the mild Galician climate, this alternative is an adequate proposal to prevent COVID infections.

Author Contributions: Conceptualization, J.A.O., M.K.N., and S.R.; methodology, J.A.O. and M.K.N.; software, S.R.; validation, S.R.; formal analysis, J.A.O. and M.K.N.; investigation, J.A.O., M.K.N., and S.R.; resources, J.A.O., M.K.N., and S.R.; data curation, J.A.O., M.K.N., and S.R.; writing—original draft preparation, J.A.O., M.K.N., and S.R.; writing—review and editing, J.A.O., M.K.N., C.M., and S.R.; visualization, J.A.O., M.K.N., C.M., and S.R.; supervision, J.A.O., M.K.N., and S.R.; project administration, J.A.O., M.K.N., and S.R.; funding acquisition, J.A.O., M.K.N., and S.R. All authors have read and agreed to the published version of the manuscript.

Funding: This research was funded by CYPE Ingenieros S.A. in their research project to reduce energy consumption in buildings and its certification, in collaboration with the University of A Coruña (Spain) (Grant No. 64900).

Acknowledgments: The authors wish to express their acknowledgement to Miguel Orosa García for his advice during the development of this work. Finally, this paper is dedicated to the memory of Antonio Fernández García (ETSNyM, University of A Coruña).

Conflicts of Interest: The authors declare no conflict of interest.

References

1. Orosa, J.A.; Oliveira, A.C. An indoor air perception method to detect fungi growth in flats. *Expert Syst. Appl.* **2012**, *39*, 3740–3746. [CrossRef]
2. ASHRAE Indications Reoupening Schools and Universities. Available online: https://www.ashrae.org/technical-resources/reopening-of-schools-and-universities (accessed on 22 September 2020).
3. REHVA CO_2 Sensor. Available online: https://www.rehva.eu/fileadmin/user_upload/REHVA_COVID-19_guidance_document_V3_03082020.pdf (accessed on 22 September 2020).
4. How Covid Travels in Classrooms. Available online: https://cse.umn.edu/college/news/new-study-explores-how-coronavirus-travels-indoors (accessed on 22 September 2020).
5. WHO. World Health Organization. Schools. Available online: https://www.who.int/publications/m/item/key-messages-and-actions-for-covid-19-prevention-and-control-in-schools (accessed on 22 September 2020).
6. Nematchoua, M.K.; Mempouo, B.; Tchinda, R.; Costa, A.M.; Orosa, J.A.; Raminosoa, C.R.R.; Mamiharijaona, R. Resource potential and energy efficiency in the buildings of Cameroon: A review. *Renew. Sustain. Energy Rev.* **2015**, *50*, 835–846.
7. Nematchoua, M.K.; Tchinda, R.; Orosa, J.A. Thermal comfort and energy consumption in modern versus traditional buildings in Cameroon: A questionnaire-based statistical study. *Appl. Energy* **2014**, *114*, 687–699. [CrossRef]
8. Peng, L.L.H.; Jim, C.Y. Green-Roof Effects on Neighborhood Microclimate and Human Thermal Sensation. *Energies* **2013**, *6*, 598–618. [CrossRef]
9. Jim, C.Y. Thermal performance of climber greenwalls: Effects of solar irradiance and orientation. *Appl. Energy* **2015**, *154*, 631–643. [CrossRef]
10. Setiawan, A.F.; Tzu-Ling, H.; Chun-Ta, T.; Chi-Ming, L. The Effects of Envelope Design Alternatives on the Energy Consumption of Residential Houses in Indonesia. *Energies* **2015**, *8*, 2788–2802. [CrossRef]
11. Parra, J.; Guardo, A.; Egusquiza, E.; Alavedra, P. Thermal Performance of Ventilated Double Skin Façades with Venetian Blinds. *Energies* **2015**, *8*, 4882–4898. [CrossRef]
12. *ASHRAE Handbook: HVAC Fundamentals*; Spanish Edition, Editorial Index; ASHRAE: Atlanta, GA, USA, 2013; pp. 1–197.
13. ISO 13790. *Thermal Performance of Buildings-Calculation of Energy Use for Space Heating and Cooling*; ISO/DIS 13790:2008; International Organization for Standardization: Geneva, Switzerland, 2008.
14. International Energy Agency (IEA). Welcome to the International Energy Agency's Energy in Buildings and Communities Programme. Available online: https://www.iea-ebc.org (accessed on 8 October 2020).
15. Haarhoff, J.; Mathews, E.H. A Monte Carlo method for thermal building simulation. *Energy Build.* **2006**, *38*, 1395–1399. [CrossRef]
16. Keirstead, J.; Shah, N. Calculating minimum energy urban layouts with mathematical programming and Monte Carlo analysis techniques. *Comput. Environ. Urban Syst.* **2011**, *35*, 368–377. [CrossRef]

17. Mavrotas, G.; Florios, K.; Vlachou, D. Energy planning of a hospital using Mathematical Programming and Monte Carlo simulation for dealing with uncertainty in the economic parameters. *Energy Convers. Manag.* **2010**, *51*, 722–731. [CrossRef]
18. Gang, W.; Wang, S.; Shan, K.; Gao, D. Impacts of cooling load calculation uncertainties on the design optimization of building cooling systems. *Energy Build.* **2015**, *94*, 1–9. [CrossRef]
19. Ren, X.; Yan, D.; Wang, C. Air-conditioning usage conditional probability model for residential buildings. *Build. Environ.* **2014**, *81*, 172–182. [CrossRef]
20. Lu, Y.; Huang, Z.; Zhang, T. Method and case study of quantitative uncertainty analysis in building energy consumption inventories. *Energy Build.* **2013**, *57*, 193–198. [CrossRef]
21. Srinivasa Rao, A.S.R. On joint Weibull probability density functions. *Appl. Math. Lett.* **2005**, *18*, 1224–1227. [CrossRef]
22. Prabhakar Murthy, D.N.; Bulmer, M.; Eccleston, J.A. Weibull model selection for reliability modelling. *Reliab. Eng. Syst. Safe* **2004**, *86*, 257–267. [CrossRef]
23. MeteoGalicia. Anuario Climatolóxico de Galicia. Consellería de Medio Ambiente. Xunta de Galicia. Available online: https://www.meteogalicia.gal/observacion/informesclima/informesIndex.action?request_locale=es (accessed on 13 October 2020).
24. Orosa, J.A.; Oliveira, A.C. Implementation of a method in EN ISO 13790 for calculating the utilisation factor taking into account different permeability levels of internal coverings. *Energy Build.* **2010**, *42*, 598–604. [CrossRef]
25. Orosa, J.A.; García-Bustelo, E.J. Permeable coverings as a sustainable solution for indoor air thermal comfort and energy saving. *Energy Edu. Sci. Technol. A* **2012**, *29*, 583–596.
26. Masdias-Bonome, A.E.; Orosa, J.A.; Vergara, D. A New Methodology for Decision-Making in Buildings Energy Optimization. *Appl. Sci.* **2020**, *10*, 4558. [CrossRef]
27. Šadauskienė, J.; Paukštys, V.; Šeduikytė, L.; Banionis, K. Impact of Air Tightness on the Evaluation of Building Energy Performance in Lithuania. *Energies* **2014**, *7*, 4972–4987. [CrossRef]
28. Fei, F.; Zhou, S.; Mai, J.D.; Li, W.J. Development of an Indoor Airflow Energy Harvesting System for Building Environment Monitoring. *Energies* **2014**, *7*, 2985–3003. [CrossRef]
29. Nematchoua, M.K.; Tchinda, R.; Orosa, J.A.; Andreasi, W.A. Effect of wall construction materials over indoor air quality in humid and hot climate. *J. Build. Eng.* **2015**, *3*, 16–23. [CrossRef]
30. Karlsson, F.; Fahlén, P. Impact of design and thermal inertia on the energy saving potential of capacity controlled heat pump heating systems. *Int. J. Refrig.* **2008**, *31*, 1094–1103. [CrossRef]
31. Mazarrón, F.R.; Cid-Falceto, J.; Cañas, I. Ground Thermal Inertia for Energy Efficient Building Design: A Case Study on Food Industry. *Energies* **2012**, *5*, 227–242. [CrossRef]
32. Orosa, J.A. Thermal inertia and ISO 13790—Have its effects on future energy consumption been fully considered? *Energy Edu. Sci. Technol.* **2012**, *29*, 339–352.
33. Orosa, J.A.; Oliveira, A.C. A field study on building inertia and its effects on indoor thermal environment. *Renew. Energy* **2012**, *37*, 89–96. [CrossRef]
34. MeteoGalicia. Xunta de Galicia. Consellería de Medio ambiente Teorritorio e Vivenda. Available online: https://www.meteogalicia.gal/datosred/infoweb/clima/informes/estacions/anuais/2019_es.pdf (accessed on 22 September 2020).

Publisher's Note: MDPI stays neutral with regard to jurisdictional claims in published maps and institutional affiliations.

applied sciences

Article

Energy Saving Strategies and On-Site Power Generation in a University Building from a Tropical Climate

Jaqueline Litardo [1,2], Massimo Palme [3,4], Rubén Hidalgo-León [1], Fernando Amoroso [1] and Guillermo Soriano [1,*]

1 Centro de Energías Renovables y Alternativas CERA, Escuela Superior Politécnica del Litoral ESPOL, Km. 30.5 Vía Perimetral, Guayaquil EC90902, Ecuador; jaqueline.litardo@polimi.it (J.L.); rhidalgo@espol.edu.ec (R.H.-L.); famoroso@espol.edu.ec (F.A.)
2 Architecture, Built Environment and Construction Engineering Department, Politecnico di Milano, Via Bonardi 9, 20133 Milano, Italy
3 Escuela de Arquitectura, Universidad Católica del Norte, Av. Angamos 610, Antofagasta 1240000, Chile; mpalme@ucn.cl
4 Centro de Investigación Tecnológica del Agua en el Desierto, Universidad Católica del Norte, Av. Angamos 610, Antofagasta 1240000, Chile
* Correspondence: gsorian@espol.edu.ec

Abstract: This paper compares the potential for building energy saving of various passive and active strategies and on-site power generation through a grid-connected solar photovoltaic system (SPVS). The case study is a student welfare unit from a university campus located in the tropical climate (Aw) of Guayaquil, Ecuador. The proposed approach aims to identify the most effective energy saving strategy for building retrofit in this climate. For this purpose, we modeled the base line of the building and proposed energy saving scenarios that were evaluated independently. All building simulations were done in OpenStudio-EnergyPlus, while the on-site power generation was carried out using the Homer PRO software. Results indicated that the incorporation of daylighting controls accounted for the highest energy savings of around 20% and 14% in total building energy consumption, and cooling loads, respectively. Also, this strategy provided a reduction of about 35% and 43% in total building energy consumption, and cooling loads, respectively, when combined with triple low-e coating glazing and active measures. On the other hand, the total annual electric energy delivered by the SPVS (output power converter) was 66,590 kWh, from where 48,497 kWh was supplied to the building while the remaining electricity was injected into the grid.

Keywords: energy savings; daylighting; photovoltaic system; EnergyPlus; Homer PRO; Net Zero Energy Buildings

Citation: Litardo, J.; Palme, M.; Hidalgo-León, R.; Amoroso, F.; Soriano, G. Energy Saving Strategies and On-Site Power Generation in a University Building from a Tropical Climate. *Appl. Sci.* **2021**, *11*, 542. https://doi.org/10.3390/app11020542

Received: 29 November 2020
Accepted: 31 December 2020
Published: 8 January 2021

Publisher's Note: MDPI stays neutral with regard to jurisdictional claims in published maps and institutional affiliations.

1. Introduction

Energy consumption in buildings is a field of extensive research worldwide. Buildings are one of the fastest growing sectors in energy consumption in the last decades [1–3], which is due to population growth, industry development, expansion of cities, and improvement of living standards [4]. The aforementioned implies a growth in the demand for thermal comfort and indoor air quality [5]. This sector has a contribution between 20–40% of the global final energy consumption [6,7], where these percentages vary according to region. For example, buildings in European Union Member States represent about 40% of total final energy consumption and 36% of carbon emissions [8]. Electricity is one of the most widely used energy sources in buildings. It is estimated that buildings consume about 60% of the global electricity [9], which has increment its use in more than 19% between 2010 and 2018 [10]. The energy consumption of buildings is mainly affected by factors such as weather conditions, building characteristics, and the operating conditions of building systems [6]. Heating, ventilation, and air-conditioning systems (HVAC) are used to provide indoor thermal comfort and their demand depends mainly on local climate, building design and internal loads. These systems are the largest consumers of energy in buildings with about 40% of it [7,11]. Also, the energy demand for HVAC is affected by the thermal properties of the building envelope (walls, doors, roofs, windows, etc.) mainly in warm climate locations [12–14]. Artificial lighting is the second most energy-consuming system in buildings, ranging from 20% to 45% [15,16]. Educational buildings are among the main consumers of energy in the commercial building sector, where in countries, such as USA, Australia and UK, the energy consumption of these buildings is between 10% and 13% [7,17]. The frequency of use and occupancy of classrooms could considerably increase the energy bills [15]. In the case of Ecuador, commercial buildings account for 20.5% of the total Ecuadorian electricity demand [18,19]. Therefore, it is necessary to look for initiatives that help to improve the efficiency in the use of energy in the buildings maintaining the levels of comfort of the users, with minimum cost and reduction of their carbon footprints.

2. Literature Review

Various energy saving strategies have been modeled and implemented in buildings to reduce their energy consumption. Some of these measures are focused on increasing the efficiency of artificial lighting and their integration with daylighting through the development of improved control systems [15,16]. The latter is considering adequate lighting levels according to the type of activity of the area to be controlled and visual comfort [20]. Although the use of daylight could be exploited in the buildings, this must be done in a technical way as there could be a possible problem with heat and solar glare [21]. Several studies have shown considerable reductions in energy consumption with the use of intelligent lighting controls. A typical Greek classroom was analyzed in [15], where the results showed that from an annual lighting primary energy consumption of 90.5 kWh/m^2, it can be reduced to 0.55 kWh/m^2. Han et al. [22] reported that the control of a 200 W LED lamp and daylight exploitation could achieve savings of 174 kWh per year taking into consideration the operating conditions of the experiment. Likewise, two control algorithms were tested for the lighting of two offices, where the experiments shown savings of up to 70% [23]. In [24], three classrooms were studied over the course of a year with various types of lighting control. Here, the authors found savings between 18% and 46%.

Previous studies have shown the potential of implementing high performance windows and shading devices in buildings to reduce their energy consumption. Kunwar et al. [25] analyzed the use of shading devices, two types of glass and lighting controls in a test room, where the authors achieved energy savings of 25.4% in cooling and 48.5% in lighting. According to [13], advanced glazing technologies can achieve annual energy savings between 0.56 kWh/m^2 and 323 kWh/m^2 per window area, where these materials are capable of handling solar heat and visible light transmitted into the building. Somasundaram et al. [26] modeled a building with the use of a new type of low-emissivity glass over existing window glass and found that this replacement could achieve energy savings of up to 9% in air conditioning usage. Marino et al. [27] stated that in warm climates when the window surface occupies more than a quarter of the wall surface, the use of shading devices is crucial for energy saving purposes. Furthermore, the authors indicated that this measure is more efficient than improving the characteristics of the building envelope. Changing the size of the window façades taking into consideration the environmental conditions and building orientation has influence on the energy consumption of the building. The Window to Wall Ratio (WWR) parameter is relationship between the size of the window surface according of its wall surface [27–29]. Alghoul et al. [30] studied the influence of the WWR and window orientation on the cooling, heating and energy consumption of a small office in the city of Tripoli, Libya, where increasing the WWR resulted in increased cooling energy consumption. In addition, the increase of windows in the façade resulted in higher energy consumption between 6–181% among all the cases studied. However in [31], the use of sunshades in a comprehensive shade in the windows of a hotel in China showed a saving of 6.5% in the annual cooling load taking a reference WWR of 0.32.

The replacement of old equipment with lower consumption ones is another option to improve the efficiency of energy use in buildings. ENERGY STAR certified equipment could reduce energy consumption by 10–50% compared to standard equipment [32]. For example, a conventional desktop computer consumes 65 W against 54 W for the same ENERGY STAR equipment. In models made to an Italian office building, Luddeni et al. [33] evaluated two equipment replacement levels of 5.4 W/m^2 and 8.1 W/m^2 using 10.8 W/m^2 as a baseline. Here, the modeling of this measure allowed energy savings of over 10%.

The use of renewable energy sources in buildings is an eco-friendly solution. Some renewable sources that can be applied in buildings are photovoltaic (PV) panel modules, geothermal heat pumps, fuel cell systems, and solar thermal collectors [34]. The use of solar panel technology is convenient for buildings because apart from reducing the dependence on grid electricity and energy bills, it could not require long distances for power transmission as solar farms [35]. In addition, this technology reduces pollution levels and acts as a heat shield when it is placed on the roof. The shading effect on the thermal performance of the roof could reduce the energy consumption for indoor air conditioning in warm climates [36].

Nearly Zero Energy Buildings (NZEBs) are based on a group of regulations, energy polices, standards, and codes, which aim to ensure a higher improvement on building energy performance and also the implementation of on-site (or near-site) energy generation from renewable sources [37]. The objectives of the NZEBs are not only limited to new buildings [38], because existing buildings can be retrofitted considering energy saving measures and renewable generation. The growth of the building sector is accelerating the use of solutions to improve efficiency, to reduce the carbon footprint and to reduce energy consumption in buildings. The European Union is promoting the NZEB as a minimum standard for the new buildings in the coming years [39]. Similarly, strategies adopted by NZEBs in hot and humid climates are related to passive and active design, exploitation of technologies such as artificial lighting and HVAC systems, other energy saving measures and renewable generation [40,41]. Despite of this, the implementation of various energy saving measures could increase construction costs considerably, mainly if NZEB targets are pursued [40].

Each energy saving strategy must be planned, designed and modelled to estimate its impact on the building performance. Many authors have used the EnergyPlus software, initially to know if the strategies could achieve NZEB objectives or in the ideal scenario, if the modelling of strategies will achieve the expected zero balance [13,31,37]. Regarding onsite electricity generation, Homer PRO software is widely used by researchers to estimate the sizing, operation, economic costs, and other parameters, of power generation systems from renewable sources [42,43].

The present study aims to evaluate the potential for energy saving of different passive and active strategies for a student welfare unit located in Guayaquil, Ecuador. The case study is surrounding by a tropical climate, with two seasons: Wet and dry. All building simulations were carried out in OpenStudio-EnergyPlus. Also, we proposed a grid-connected solar photovoltaic system (SPVS) to reduce the purchase of electricity from the grid. The expected results would determine some guidelines for building retrofits in this and related climates. Also, the study can contribute to generalize energy saving strategies for tropical cities, Aw, according to the Köppen-Geiger classification [44], and improve the sustainable building design in these climates. This is a relevant issue since more than 33% of the global population lives in the tropics, and in Ecuador more than 50% of its total population [45].

3. Materials and Methods

The methodology has been divided into five subsections. Section 3.1 addresses location and climate conditions of the surroundings of the studied building. Section 3.2 describes the geometry, building materials, and the operation schedules of the modeled building. Section 3.3 describes some details about OpenStudio-EnergyPlus simulations. Section 3.4

defines energy saving strategies to be proved for the case study. Finally, Section 3.5 encompasses the on-site power generation through a photovoltaic system.

3.1. Location and Climate Conditions

The city of Guayaquil (2°11′21.89″ S, 79°53′20.64″ W) is the largest and the most populated city of Ecuador, with an estimated area of 2494 km^2 and about 2.72 million people in 2020, based on projections of the National Census Institute (INEC) [46]. Guayaquil has a tropical climate, corresponding to the Aw group from the Köppen-Geiger classification [44], with two well-defined seasons: wet (January–April) and dry (May–December). The climate conditions of Guayaquil are depicted in Figure 1. The monthly average temperatures vary between 23.4 to 26.5 °C. The monthly minimum temperatures are above 18.0 °C, while the monthly maximum temperatures are below 33.2 °C. The relative humidity values are higher during the wet season, while the annual average is about 70%. The monthly average global radiations are between the 3.5 and 5.3 kWh/m^2.

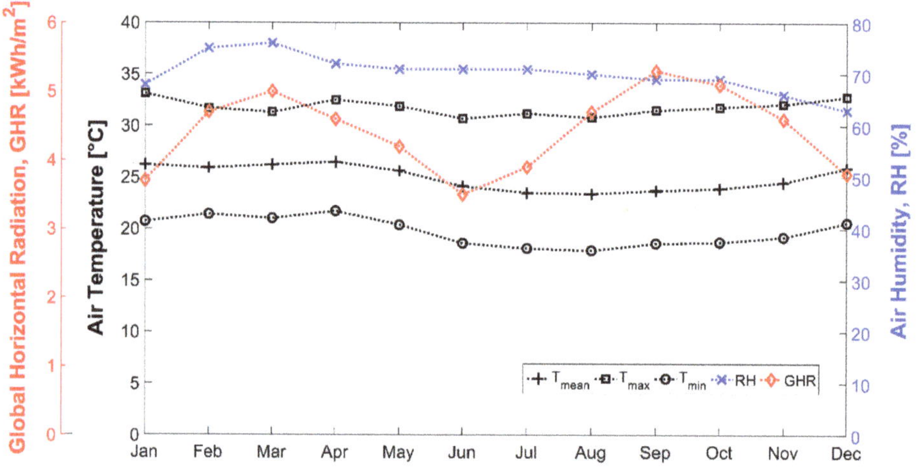

Figure 1. Climate conditions in Guayaquil, Ecuador (authors elaboration based on data from Meteonorm [47]).

Figure 2 shows the distribution of the annual climate conditions of Guayaquil (temperature and humidity) plotted on the psychrometric chart based on a Typical Meteorological Year (TMY) file. As can be observed, during almost all the year, the city presents high outdoor temperatures and humidity. It means that around 86% of the year people could experiment indoor discomfort. Therefore, efficient air-conditioning systems and/or passive energy saving strategies are required to reduce the energy demand for cooling in these climates. Some strategies suggested by the Climate Consultant v. 6.0 tool [48] for the studied climate are dehumidification, sustainable cooling, and solar shading.

3.2. Building Description

The building under study is the Student Welfare Unit of ESPOL Polytechnic University, whose campus is in Guayaquil, Ecuador (Lat. 2° 8′34.61″ S, Long. 79°58′1.76″ O) (Figure 3). It is a two-story building formed by conventional local building materials, based on masonry construction [14]. Table 1 summarizes the physical and thermal properties of the building materials that form the studied building envelope. The entire building has a total floor area of about 1086 m^2 and a floor to floor height of 4.5 m. This building has a 35% window-to-wall on the north wall, 38% on the south wall, 33% on the east wall, and 23% on the west wall.

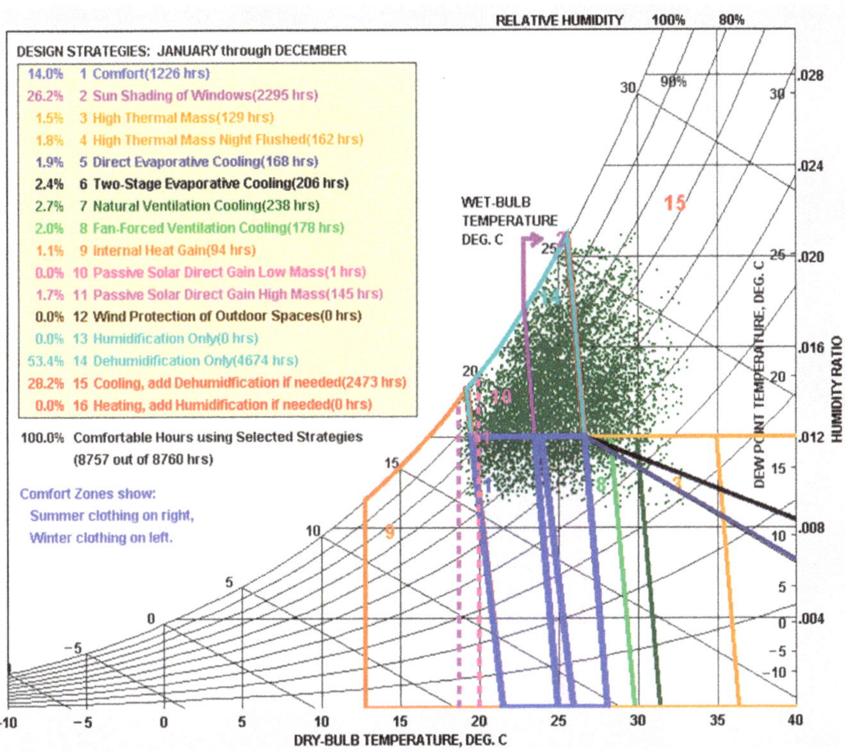

Figure 2. Hourly outdoor temperature and humidity in Guayaquil, Ecuador (obtained from Climate Consultant [48]).

Figure 3. Student Welfare Unit building of ESPOL Polytechnic University (from [49]).

Table 1. Physical and thermal properties of envelope building materials (adapted from [14]).

Construction	Density [kg/m^3]	Thermal Conductivity [W/m-K]	Specific Heat [J/kg-K]	U-Value [W/m^2-K]
Roof				
Steel sheet (1 mm)	7800	50	450	
Heavy weight concrete (30 cm)	2240	1.31	837	
Wall				
Hollow concrete block (9 cm)	1600	0.47	1000	
Plaster (1 cm)	800	0.37	340	
Floor				
Heavy weight concrete (10 cm)	2240	1.31	837	
Acoustic tile (2 cm)	368	0.06	59	
Windows				
Clear glass (6 mm)				
Metal frame				5.78

The building has a double use: it works as an office building and as a medical center for students and the rest of the University staff. The building operates from 8:00 to 16:30 during weekdays.

The internal heat gains of the building consist of internal (appliances, lighting) and conditioned loads (direct expansion mini splits). Table 2 summarizes the overall installed power by end-use. All the internal and conditioned loads of the building work with electricity.

Table 2. Building internal loads: installed power by end-use (from [49]).

System	Description	Installed Power [W]
Lighting	LED and saving lighting	11,974
Electric equipment	Appliances	25,716
Air-conditioning	Direct expansion mini and floor/ceiling splits	170,567 (581,999 BTU/h)

Also, the studied building has an occupancy of around 30 permanent workers, all of them performing light office activities. Therefore, the activity level value was set at 100 W/person.

The energy tariff for the building is USD$0.06/kWh from 07:00 to 22:00; the rest of the hours the electricity cost is USD$0.05/kWh. Likewise, the tariff for monthly peak demand is USD$1.57/kW. The city's public electricity company established these values for 2020 [50].

3.3. Building Energy Model in EnergyPlus

We used the OpenStudio v. 2.7 tool and EnergyPlus v. 9.1 software for the simulations. As a first step, we modeled the baseline of the building, i.e., its actual situation in terms of energy (Figure 4). For this, we considered all internal loads from Table 2, and people occupancy.

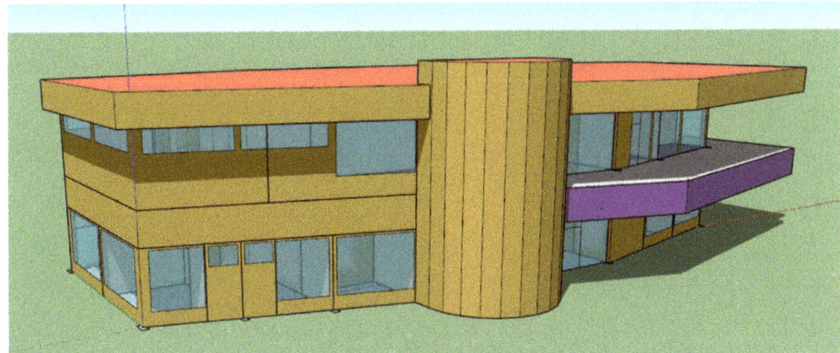

Figure 4. Student Welfare Unit building model in EnergyPlus (authors elaboration).

We established the schedules of the building based on the information provided by the users. We fixed the infiltration rates in 0.54 air changes per hour (ACH) [18]. Air-conditioners were modeled with a standard COP of 3, considering that the systems installed in the case study are obsolete.

3.4. Energy Saving Scenarios

Previous studies have defined several strategies to improve energy use in buildings [33]. The optimal selection of these will depend on the final use of the building, the needs of the users (environmental comfort), the investment costs, and the climate conditions of the site. In general, energy saving measures can be divided into two large groups: passive and active strategies. Passive strategies are mainly focused on improving the thermal properties of the envelope components and/or taking advantage of the climate and surroundings to reduce the energy demand of the building. In contrast, active strategies focus directly on improving the energy efficiency of the building's internal loads and/or on the implementation of smart controls.

In this section, we propose different energy saving scenarios to analyze the feasibility of incorporating them in the retrofit of the case study. All scenarios were tested independently.

3.4.1. Integration of Daylight with Artificial Lighting

In order to take advantage of the daylight levels around the studied building, we proposed a dimming control.

Figure 5 depicts the schematic of the dimming control scenario modeled in EnergyPlus. This system uses a photosensor to monitor the lighting level (in lux) in a reference point of the studied zone. As the main purpose of these systems is to maintain the desired lighting levels within the zone, they adjust the output power of the artificial lighting systems based on the available daylight levels. We used "SplitFlux" as the daylighting method, with a minimum input power fraction and minimum light output fraction of 10% in both cases. We also defined the illuminance setpoint at each reference point as 500 lux.

3.4.2. Window to Wall Ratio (WWR-20%)

Since the optimal WWR for daylighting and natural ventilation is around the 30–40% and considering the external view as an important factor for the environmental comfort of the users, we proposed a reduction of the WWR from the base case to 20% in all façades. For this purpose, we used the "Resize existing windows to match a given WWR" measure from NREL's Building Component Library (BCL) available for OpenStudio [51]. This measure reduces the dimensions of each window around its centroid and it only works in case of WWR reduction. The result of applying this measure provides a clearer perspective about its potential against the potential of daylighting control strategy. Therefore, from this analysis we would be able to know if it is recommended to keep the current WWR to take advantage of the daylighting availability or if it is better to reduce it without considering the incorporation of daylighting strategies in this building.

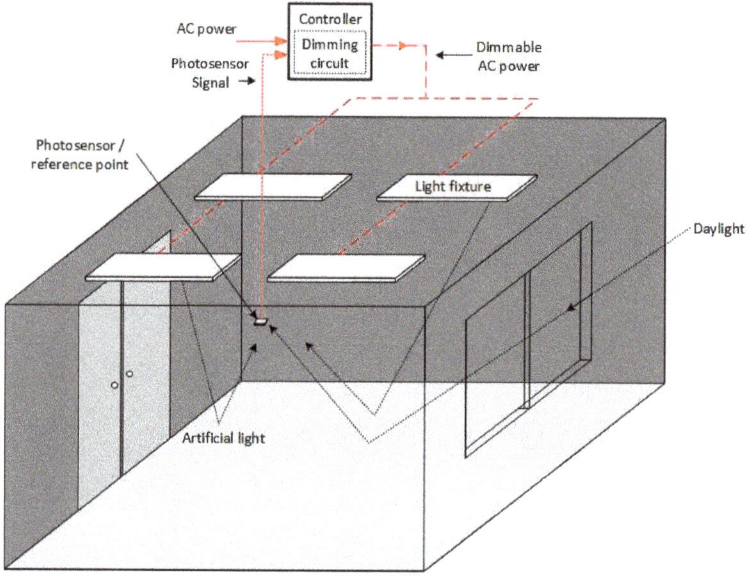

Figure 5. Schematic of a zone using a dimming control system from the integration of daylight and artificial lighting (authors elaboration).

3.4.3. Static Solar Shading

We employed static shading devices on the windows of the building in horizontal (overhangs) and vertical fins configurations, as shown in Figure 6. These configurations aim to reduce the penetration of sunlight into the building through the windows [31]. These structures were placed on the windows of all façades in the building model. The height and width of these structures are adapted to the dimensions of the windows and the depth was set at 60 cm in all windows.

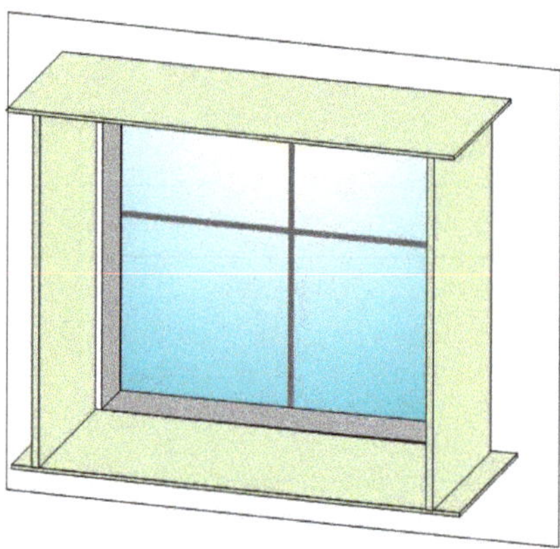

Figure 6. Horizontal and vertical shadings (adapted from [31]).

3.4.4. High Performance Windows (TrpLoE)

Static windows with triple glazing were considered in the analyses, as shown in Figure 7. This retrofit window has a thickness of 44 mm, where the glasses are in layers 1–2, 3–4 and 5–6. It has two air gaps filled with argon gas (2–3 and 4–5), as well as two low-e coating films (2 and 5). The thickness of each air gap with argon is 13 mm and 6 mm for each glazing. These retrofit windows were modeled in all façades. We proposed windows with lower thermal transmittance values (U-value = 0.785 W/m^2-K) to reduce the solar gains of the building, while admitting the entrance of daylight into the indoor spaces (visible light transmittance, τ = 0.66), with a low solar heat gain coefficient (SHGC = 0.474) to avoid overheating. Single pane windows, which are common in the façades of Guayaquil's buildings, are responsible for increasing the energy demand for cooling due to the higher solar gains. Poor window design will result in significant impacts on thermal and visual comfort and increased energy demand for cooling, and consequently, these will increase energy bills and CO_2 emissions.

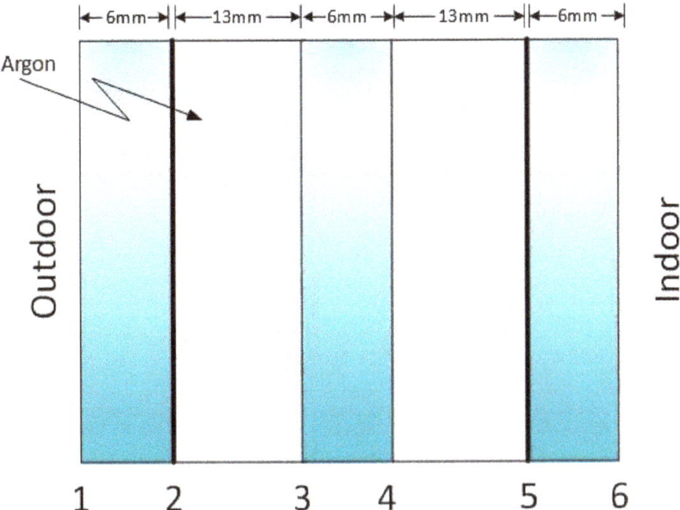

Figure 7. Triple glazing low-e coating window filled with argon, TrpLoE (authors elaboration).

3.4.5. Active Measures for Energy Saving

We analyzed three actives measures, which are: changing the cooling set point of the thermal zones, increasing the coefficient of performance (COP) of the air-conditioners, and replacing the standard electric equipment with others with ENERGY STAR certification.

In the studied case, the set points of the air-conditioners are typically fixed at temperatures between 21–22 °C. For energy saving purposes, we changed the set point temperature to 24 °C, in accordance with [45] to maintain an adequate level of comfort for all users. This measure is intended to establish a single temperature, which will reduce the excessive cooling of areas according to the needs of certain users.

For this scenario, we also considered the replacement of air-conditioners with others of higher coefficient of performance (COP). To accomplish this, we followed the recommendations of the ASHRAE 90.1-2019 Standard [52] for the U. S. Department of Energy (DOE) minimum energy efficiency requirements for HVAC systems and changed the COP values of each one in OpenStudio. Table 3 summarizes the cooling capacity ranges of the air-conditioners installed in the case study and their corresponding minimum COP based on DOE recommendations.

Table 3. Cooling capacity ranges and DOE minimum energy efficiency requirements for air conditioners (extracted from [52]).

Cooling Capacity [BTU/h]	COP
<65,000	3.45 (SEER 14)
≥65,000 and <135,000	3.55 (SEER 14.6)

The replacement of electric standard office equipment with similar but less energy-consuming equipment was another energy-saving measure proposed in this study. We proceeded to replace office equipment such as computers, printers, and copy machines by similar equipment but with ENERGY STAR certification [32,33]. This is equivalent to a reduction of about 20% in the installed power for appliances. This measure maintained the same operation profiles of the base case. The use of high-performance lighting systems was not in the scope of this study since it is considered that the lighting systems of the base case are LED.

3.4.6. Combined Scenarios

We also proved two combined scenarios that were selected at our discretion:

1. Scenario 1: Daylighting control + TrpLoE + Active measures
2. Scenario 2: Shading + TrpLoE + Actives measures

We considered the active measures in both scenarios since these strategies are the directly related to the reduction in internal loads. On the other hand, we chose to evaluate the TrpLoe windows in both cases to answer one key question: Is it better to take advantage of the solar radiation coming into the building, or is it preferable to block it partially, considering the studied climate?

3.5. On-Site Energy Generation

The building under study has a nearby total area of more than 2000 m^2, free of obstacles, where PV modules could be installed without any inconvenience in case of implementation. For the purpose of the present work, we proposed to use between 20% to 25% of the total available area. Taking into account the size and power of the selected PV module and the target area, we considered a grid connected solar PV system (SPVS) of 50 kWp for on-site energy generation.

The SPVS aims to reduce the use of the building's electrical energy from the electric power grid. In the implementation of a SPVS, its investment costs could be much higher than the execution of the other energy saving measures in the building, depending mainly on its size [33]. In the simulation it is assumed that the surfaces of the PV modules are facing north with a tilt angle similar to the latitude of the site. These last conditions are ideal for extracting the most significant PV energy in SPVSs, located in the southern hemisphere [53]. DC/AC power converters sends the power generated by the PV panels to the electric power grid, which can operate in parallel to the grid without any inconvenience.

The simulations of this system were carried out in the HOMER Pro software [54], taking into consideration the electrical load modeled in EnergyPlus. The governing equation to calculate the output power of each system is the following:

$$P_{out,PVsyst} = Y_{PVsyst} \times f_{PVsyst}\left(\frac{G_T}{G_{T,PVsyst}}\right)\left(1 + \alpha_p\left(T_c - T_{c,PVsyst}\right)\right), \quad (1)$$

where Y_{PVsyst} is the nominal capacity of the PV system, f_{PVsyst} is the derating factor of the PV system, G_T is the incident radiation on the PV system [kW/m^2], $G_{T,PVsyst}$ is the irradiation under standard test conditions [1 kW/m^2], α_p is the temperature coefficient of the power [%/°C], T_c is the cell temperature of the PV system [°C], and $T_{c,PVsyst}$ is the temperature of the PV system under standard test conditions [25 °C].

The PV module considered for the calculations was 360 Wp, monocrystalline type, with a surface area of 1.77 m^2 [55]. The cost for this module in the Ecuadorian market is around USD$220.00 per unit with a lifetime of 25 years [56]. In addition, we considered a derating factor of 90%, where 10% represents the losses such as dust on the surface of the panels, wiring, shading, and others [57]. Likewise, the cost of the 6 kW DC/AC power converter was USD$5,000.00, with an efficiency of 95%. The replacement costs of these two equipment are not considered in this simulation.

Figure 8 shows the configuration of SPVS, which was simulated in Homer Pro. It is composed of PV panels, DC/AC power converters, electric power grid and the electrical load of the building.

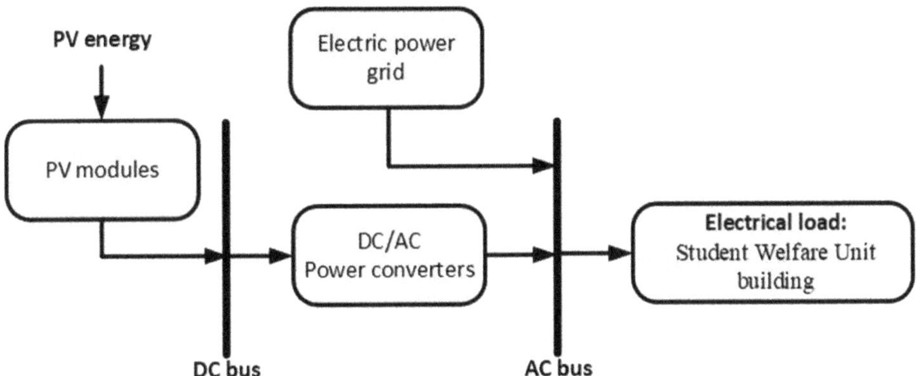

Figure 8. Configuration of grid-connected solar PV system of 50 kWp to be simulated in Homer PRO (authors elaboration).

4. Results

In order to provide a better understanding of this section, it has been structured as follows. The first subsection encompasses the estimated results of the annual energy consumption of the building and its energy consumption by end-use. These results represent the baseline of the model. The second subsection presents the results of the energy saving strategies and analyzes their feasibility to improve the energy use in the studied building. The third subsection presents the results of the combined energy saving scenarios. The forth subsection regards to the costs related with each of the strategies. Finally, the fifth section shows the results from the SPVS and its related costs.

4.1. Baseline

We estimated the energy consumption of the Student Welfare Unit model related to the annual operational activities of the building. Considering all internal gains and conditioned loads, the estimated annual energy consumption resulted in 97,958 kWh. The estimated Energy Use Index (EUI) was 90.17 kWh/m^2-year. The 44% of the total energy consumption corresponded to the use of HVAC systems. Lighting and electric equipment occupied 31% and 25% of the total, respectively. The results are consistent with analysis conducted in similar buildings of the Coastal Region of Ecuador [58,59].

Figure 9 shows the estimated monthly energy consumption by end-use. As can be observed, the energy consumption exhibited a seasonal behavior being higher during the months of the wet season and lower during the months of the dry season. This is an expected result since outdoor temperatures during the wet season are higher causing heat gains through the envelope and through infiltration to be more significant than during the dry season.

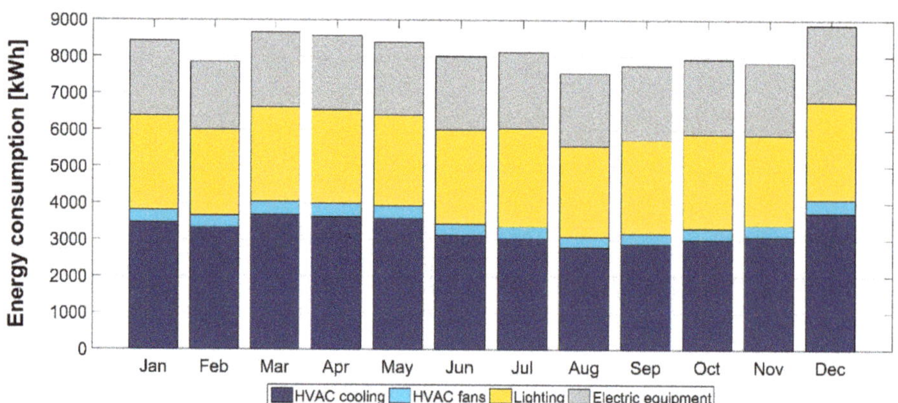

Figure 9. Estimated monthly energy consumption by end-use (authors elaboration).

Since energy consumption for cooling and lighting are the largest, measures should be taken to reduce electricity consumption mainly in these sectors, without affecting the comfort levels of the indoor environment (thermal, visual, and others). Regarding electric equipment, their replacement with their equivalent ENERGY STAR could lead to a relevant reduction in the annual energy consumption of the building. Also, energy efficiency policies could be addressed to improve the energy use in this type of buildings. The introduction of common dinning spaces and user training are other initiatives that can contribute to reduce the energy consumption and consequently, the carbon footprint of the building.

4.2. Energy Saving Scenarios

Figure 10 shows the results from the simulations of the energy saving scenarios proposed for the case study. We analyzed the savings strategies in the figure separately. As can be seen, daylighting control strategy had a highest impact on the total energy consumption in comparison with the other studied strategies. It provided a higher reduction in the annual building consumption by decreasing the energy for lighting in 42% compared with the base case. This strategy exhibited a EUI of 72.09 kWh/m^2-year. Next, active measures were in second place, providing the highest reduction in energy for HVAC systems and equipment. This is an expected result since active measures directly increased the energy efficiency of the HVAC systems, and also, the replacement of obsolete equipment with their ENERGY STAR equivalent reduced the final energy consumption for equipment in 15%, which can increase if it is applied to other type of buildings such as offices. The resultant EUI of this strategy was 77.07 kWh/m^2-year. The other saving strategies showed similar percentages of reduction. The main reason why the daylighting strategy had the highest impact among all strategies is due to the exploitation of daylight in the lighting of the building areas. This contribution of daylight to artificial lighting reduces energy consumption considerably compared to the case where there were not dimmable control systems on the light fixtures. Also, the application of all strategies reduces the energy demand for cooling of the building so that there was a reduction in cooling compared to the base case. Likewise, there was a slight reduction in the use of HVAC fans. These results help to estimate how much electrical energy can be saved by implementing these five strategies.

Figure 11 shows the energy saving percentages and the impact of each energy saving strategy on the total building electricity consumption and cooling. As can be observed, WWR-20% showed the smallest reduction in the annual total energy consumption (only 5.6%) but provided a reduction of about 12% in cooling loads. The solar shading and TrpLoE strategies showed reductions in energy for cooling of around 15–16%. The implementation of the WWR and Daylighting strategies should be carefully analyzed in building modeling. This is because while window reduction also reduces internal heat gains, this measure primarily affects the input of daylight and natural ventilation. As can be seen, the daylighting strategy showed a reduction of 20%, being the one that provided more savings in annual energy consumption. Despite daylighting controls decrease the energy for cooling in 14.3%, active measures provided a higher saving in cooling load of about 24.2%. In the case of the building under study, the daylighting strategy would be the ideal choice (between the two strategies mentioned) that would represent a considerable energy savings, in this particular case in electricity, and at the same time, a decrease in the energy bills.

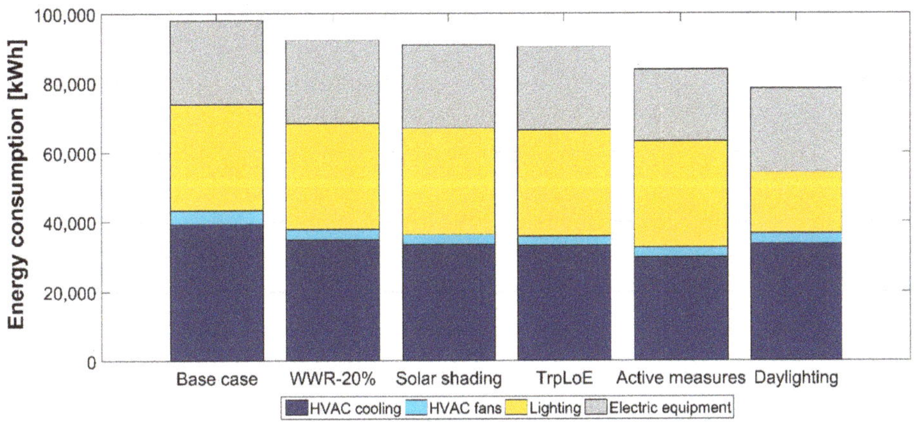

Figure 10. Annual energy consumption by end-use estimated for each proposed scenario (author's elaboration).

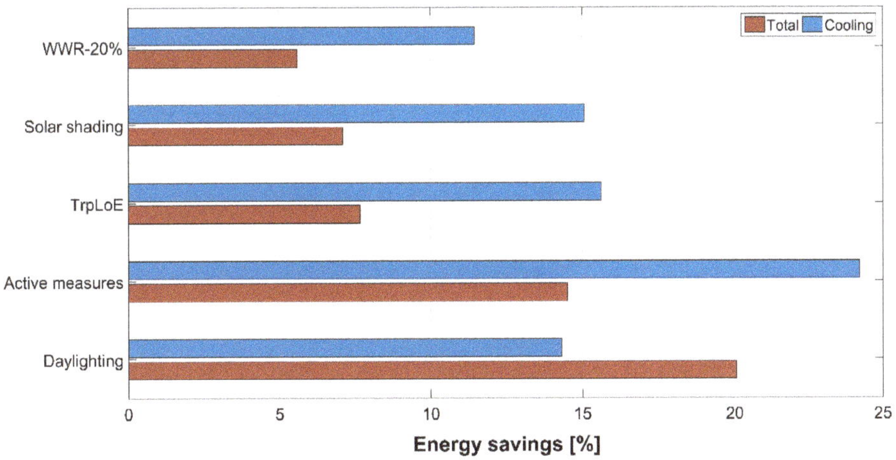

Figure 11. Annual energy saving percentages for cooling and total loads obtained from each proposed scenario (authors elaboration).

4.3. Combined Scenarios

In this section, we simulated two scenarios consisting of two groups of strategies. Scenario 1 consists of the application of solar shading, TrpLoE and active measures, while Scenario 2 consists of the application of daylighting, TrpLoE and active measures in the building model. Results from the simulations of these scenarios can be observed in Figure 12. In the Figure 12a, scenario 1 showed an annual energy consumption of 75,203 kWh

(EUI = 69.22 kWh/m^2-year) and scenario 2 of 63,713 kWh (EUI = 58.64 kWh/m^2-year). Figure 12b showed the percentage savings in total energy consumption and in cooling in the modeling of the two scenarios. Scenario 1 reduced total energy consumption by 35% and scenario 2 by 23% compared to the base case. The impact of both scenarios on building cooling is about 43%.

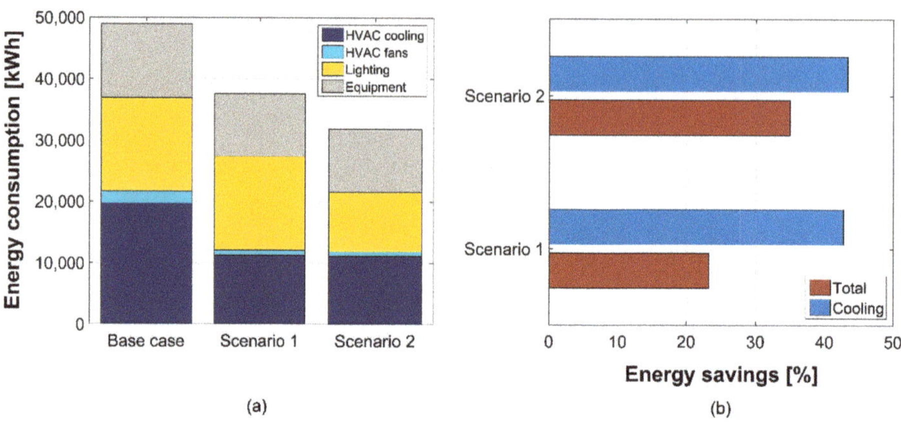

Figure 12. Estimated annual; (**a**) energy consumption; and (**b**) savings obtained from combined scenarios (authors elaboration).

4.4. Investment Cost of the Proposed Energy Saving Scenarios

In this section, we present the investment costs related to each of the proposed energy saving scenarios. Table 4 lists the approximated costs taking into account the suggested costs from the Ecuadorian construction sector [60,61]. As can be observed, the costs related to implementation of high performance windows are the highest and also, it presented the highest payback (more than 100 years). On the other hand, the installation of daylighting controls provided the lowest costs and a payback of more than 2 years. For this, daylighting solutions could be considered as the most cost-effective strategy to provide energy savings in this building according to the present analysis. Also, changing the cooling setpoint of the building should not be ignored since this strategy does not present an investment cost. Other strategies as WWR-20% and the use of static shadings presented paybacks of less than 50 years. The implementation of this strategies should be carefully analyzed, especially due to the cost of the energy in this building, which is subsidized. Therefore, this decision would probably lie on the availability of external investors.

Table 4. Investment costs of the proposed energy saving scenarios (authors elaboration based on costs established in [60,61]). All costs are presented in USD$.

Energy Saving Scenario	Investment Cost	Unit	Total	Payback
Daylighting controls				
Dimming controller Photosensor	From $100.00/unit From $60.00/unit	20 20	From $3200.00	2.34 years
Window to wall ratio reduction				
WWR-20% Windows desmounting Cement filling New windows installation	From $4.25/m^2 From $10.00/m^2 From $31.00/m^2	310 m^2 181 m^2 129 m^2	From $7126.50	18.7 years

Table 4. *Cont.*

Energy Saving Scenario	Investment Cost	Unit	Total	Payback
Static shading devices				
Overhangs and vertical fins from aluminium	From \$118.76/m^2	211 m^2	From \$25,058.36	47.16 years
TrpLoE				
Triple glazing (6 mm) with double low emissivity films and aluminum profile	From \$174.80/m^2	310 m^2	From \$54,188.00	103.8 years
Active measures				
Changing the cooling setpoint temperature	-	-	-	63.7 years
Replacement of obsolete air-conditioners with their higher efficiency equivalent	Mini split 12,000 BTU/h: from \$500.00/unit Mini split 18,000 BTU/h: from \$700.00/unit Floor/ceiling split 36,000 BTU/h: from \$950.00/unit Floor/ceiling split 60,000 BTU/h: from \$1200.00/unit	10 7 6 2	From \$18,000.00	
Replacemet of obsolete electric equipment with their equivalent ENERGY STAR	Computer: from \$1100.00/unit Copy machine: from \$1200.00/unit Printer: from \$200.00/unit	31 9 1	From \$45,100.00	

4.5. On-Site Power Generation

Figure 13 shows the estimated monthly electricity generation of the SPVS of 50 kWp projected by HOMER Pro. The electrical load of the building was 97,958 kWh/year, whose monthly distribution can be seen in Figure 5. The total annual electric energy delivered by SPVS was 66,590 kWh where 48,497 kWh served to feed the building and the remaining 18,093 kWh were introduced into the electric power grid. This last amount is the excess energy produced by the SPVS when the building's electrical load is already covered [62]. The energy purchased from the grid during one year for the building's operation was 47,080 kWh. The month of July showed the minimum reduction of 44.20% in the use of energy purchased from the grid compared with the baseline modeling. October showed the highest reduction with 59.96%. The average monthly reduction was 51.98%.

Figure 14 shows the distributions of the annual hourly electrical power. As can be seen in the figure, the distribution of the power from the SPVS (output power converter) and electric power grid to cover the electrical load can be clearly observed. It should be emphasized that the operation of the SPVS is given according to the following cases. First, when there is a high solar resource and certain power consumption of the building at a specific time during daytime, the power of the converter could exceed this consumption and the surplus is introduced into the grid. The highest levels of power injection to the grid take place between 11:00 and 17:00, given that the electrical load is completely covered. In contrast, if there is a low solar resource and certain power consumption of the building in a certain time during daytime, the power of the converter is completely injected into the building and the difference to complete the load is taken from the grid. During the nighttime, the power consumption of the building is completely taken from the grid.

Furthermore, since the building does not operate on weekends, the power delivered by the SPVS is almost entirely injected into the grid. The results show that there are a significant number of hours of daytime per year that present the first case of operation.

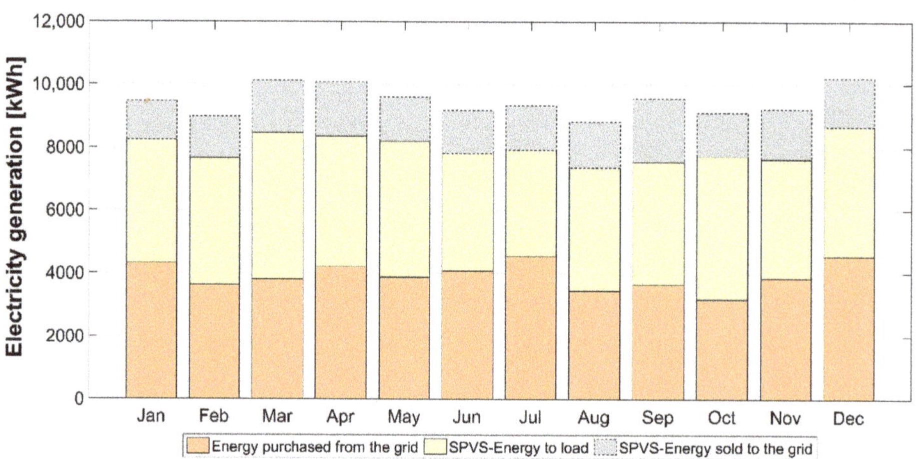

Figure 13. Estimated monthly electricity generation for a 50 kW PV system (authors elaboration).

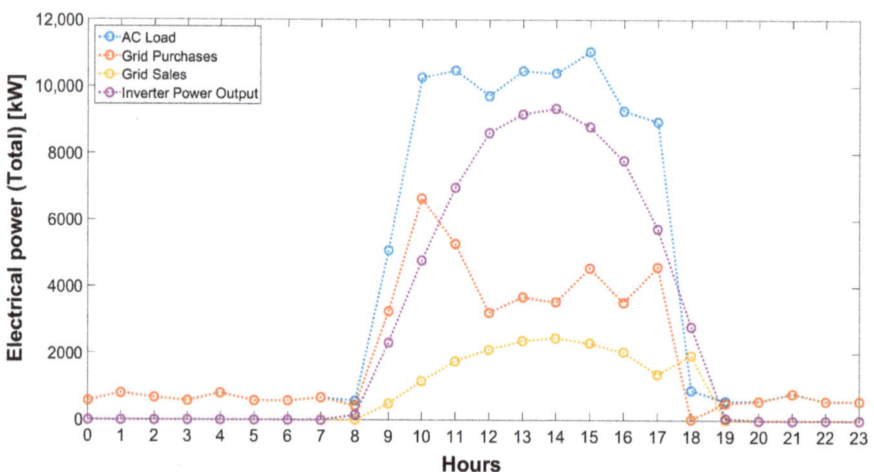

Figure 14. Distributions of the total annual electrical power of the load, SPVS, and electric power grid (author's elaboration).

The initial capital cost of the SPVS was USD$72,222.00 with an operation and maintenance cost of USD$3211.00. This last cost is related to the payment of energy purchased from the grid. In the present work, the SPVS was analyzed for one year of operation and the costs of assembly and installation of the system had not been considered.

5. Discussion

Energy savings strategies based on windows retrofits are one of the most expensive according to literature [33,63]. In our analysis, this strategy provided a reduction of about 16% and 8% in cooling loads, and total building consumption, respectively. The use of daylighting controls compared with the above exhibited a higher potential for savings that can lead to reductions of around 20% or more, depending on the control technology [64]. In equatorial zones, as in the case of Guayaquil, daylighting strategies can be very effective as stated in literature [65,66] since more direct solar radiation is received in this areas. This alternative could also present superior cost benefits when compared with the implementation of higher performance windows (doble reflective or triple pane low-e windows) or the replacement of obsolete equipment with more energy-efficient devices. The introduction of higher COP air-conditioners in old buildings could represent an important investment that should not be neglected, even when this measure could allow for energy savings ranging from 6 to more than 30% for different SEER values [67]. Therefore, this should be rethought for being used in new building projects and prioritize other low-cost measures for retrofitting purposes.

For minimizing the costs of implementing these measures in retrofits, some zero-cost strategies should be considered to improve energy use in buildings. Papadopoulos et al. (2019) [68] showed an example of this by changing the thermostat setpoint without compromising the occupants thermal comfort levels. When varying the cooling occupied setpoint between 24.3 to 26.9 °C, they found reductions up to 60% in HVAC loads depending on the climate. In this study, we included this strategy between the active measures and obtained that around 1/3 of the reduction provided by this strategy corresponded to the case when thermostat was set at 24 °C. This is equivalent to a decrease of 4% in the annual building electricity bills, which can be used to partially meet the implementation costs of other strategies. Despite the benefits of setting a higher static temperature for cooling control, some studies suggest that adaptive setpoints for summer and winter could reduce the building annual energy consumption even more [69,70]. In the case of Guayaquil, it is required to deepen this analysis to determine its effectiveness since the climate of this city could be a challenge for this approach.

Other strategies, such as the installation of static shading devices have been widely studied for researchers worldwide and despite their limitations [71], their use in hot and humid climates could account for reductions between 10% and even more than 20% in building cooling loads, depending on the shading design [72,73]. Even though our results are in accordance with this range, more effort should be devoted to exploring the optimal design of these element to decrease the energy demand for cooling in Guayaquil buildings, considering its climate, the type of building, and also, the amount of solar radiation on each façade. Full shading in all façades is recommended for tropical climate cities as Guayaquil [72,74]. However, to determine the best shading configuration for each façade could allow for relevant reductions in cooling, while decreasing the costs related to the application of this strategy.

The power of SPVS does not reduce the energy consumption of the building like other energy saving strategies. This power is consumed by the building during the daytime on-site and the surpluses sent to the power grid [33]. According to the legislation in Ecuador, these surpluses could generate an energy credit in favor of the consumer but with no money back from the electric company [75]. Guayaquil is a city that has a considerable solar resource [76], where the implementation of SPVSs could considerably reduce the electricity bills and the carbon footprint of buildings. However, the expansion of these systems is limited by the subsidies that have certain countries on electricity tariffs [77], as in the case of Ecuador.

The obtained results from EnergyPlus simulations could present uncertainties of around ±10% [78,79], which is in accordance with the ASHRAE Guideline 14 [80]. These uncertainties depend on external errors related to weather data, schedules, and others, but also, internal errors such as the difference between actual HVAC and its simplified model in the simulation, and the inaccuracies of the mathematical models [81].

Overall, this study showed the potential for energy savings of various passive and actives measures applied in a university building from Guayaquil, Ecuador. From these results, some research questions emerged that could be addressed in future works:

- What could be the best shading configurations to be applied in Guayaquil taking into account the incident solar radiation in each façade?

- What could be the best cooling set point to save energy considering the thermal comfort of the occupants in these type of buildings?
    - Should this cooling set point change depending on the season (wet/dry)?
- What could be the best cooling strategy to provide energy savings and reduce the building's carbon footprint without compromising indoor thermal comfort in this climate?
- Apart from the implementation of daylighting controls, what could be another low-payback alternative to reduce energy in this climate?

6. Conclusions

This paper analyzed the potential for the implementation of some energy saving strategies in building retrofits from the tropics. The base case, which was a student welfare unit from an Ecuadorian university campus, and all the proposed energy saving strategies were modeled in OpenStudio-EnergyPlus. This study also evaluated the opportunities of the implementation of a SPVS in the studied building, which was simulated in Homer PRO software.

The analysis of the proposed strategies suggested energy savings between 10 and 25% in cooling loads and between 5 and 20% in the annual building consumption when were evaluated independently. Whereas the combination of some strategies resulted in savings of around 43% in cooling and from 23% to 35% in annual building consumption.

Daylighting control strategy was the one that provided the more energy savings independently and even when combined with other strategies such as TrpLoE windows and active measures. The results suggested that it would be more cost-effective to use daylighting strategies in building retrofits located in hot and humid climates.

The use of SPVS reduces the carbon footprint and the electricity bills of the building; however, its implementation is subjected to local regulations and the total costs of the project.

The combination of the energy saving strategies with the proposed PV system could approach the case study into a NZEB.

7. Limitations of the Study

This study presents a model of a real building located in the tropical city of Guayaquil, Ecuador. Its limitations lie particularly in the use of TMY weather files and in the power of the simulation tools. Also, the energy saving scenarios, proved in this work, are based on some strategies, which have been previously recommended by the research community for hot and humid climates.

Despite these limitations, results were congruent with the related literature and can be used by architects, engineers and/or designers for the construction or retrofit of local buildings or related.

Author Contributions: Conceptualization, J.L. and R.H.-L.; methodology, J.L., M.P. and G.S.; software, J.L., R.H.-L., F.A.; validation, J.L., M.P. and G.S.; formal analysis, J.L. and R.H.-L.; investigation, J.L., R.H.-L. and F.A.; resources, G.S. and M.P.; data curation, J.L. and R.H.-L.; writing—original draft preparation, J.L., R.H.-L. and F.A.; writing—review and editing, M.P. and G.S.; visualization, J.L.; supervision, M.P. and G.S.; project administration, M.P. and G.S. All authors have read and agreed to the published version of the manuscript.

Funding: This research received no external funding.

Informed Consent Statement: Not applicable.

Data Availability Statement: Data sharing is not applicable to this article.

Acknowledgments: The authors would like to express their acknowledgements to ESPOL's Research Deanery for supporting the funding for this study.

Conflicts of Interest: The authors declare no conflict of interest.

References

1. IEA. *Electricity Consumption, World 1990–2018;* IEA: Paris, France, 2019.
2. GlobalABC; UNE; IEA. *Global Status Report for Buildings and Construction: Towards a Zero Emissions, Efficient and Resilient Buildings and Construction Sector;* UN Environment and IEA: Paris, France, 2019.
3. IEA. *International Energy Outlook 2019 with Projections to 2050;* IEA: Paris, France, 2019.
4. Hidalgo-Leon, R.; Urquizo, J.; Macias, J.; Siguenza, D.; Singh, P.; Wu, J.; Soriano, G. Energy Harvesting Technologies: Analysis of Their Potential for Supplying Power to Sensors in Buildings. In Proceedings of the 2018 IEEE 3rd Ecuador Technical Chapters Meeting, Cuenca, Ecuador, 15–19 October 2018.
5. Lu, T.; Lü, X.; Viljanen, M. A Novel and Dynamic Demand-Controlled Ventilation Strategy for CO2 Control and Energy Saving in Buildings. *Energy Build.* **2011**, *43*, 2499–2508. [CrossRef]
6. Li, X.; Zhou, Y.; Yu, S.; Jia, G.; Li, H.; Li, W. Urban Heat Island Impacts on Building Energy Consumption: A Review of Approaches and Findings. *Energy* **2019**, *174*, 407–419. [CrossRef]
7. Jafarinejad, T.; Erfani, A.; Fathi, A.; Shafii, M.B. Bi-Level Energy-Efficient Occupancy Profile Optimization Integrated with Demand-Driven Control Strategy: University Building Energy Saving. *Sustain. Cities Soc.* **2019**, *48*, 101539. [CrossRef]
8. Dascalaki, E.G.; Balaras, C.A.; Kontoyiannidis, S.; Droutsa, K.G. Modeling Energy Refurbishment Scenarios for the Hellenic Residential Building Stock towards the 2020 & 2030 Targets. *Energy Build.* **2016**, *132*, 74–90.
9. Manic, M.; Wijayasekara, D.; Amarasinghe, K.; Rodriguez-Andina, J.J. Building Energy Management Systems: The Age of Intelligent and Adaptive Buildings. *IEEE Ind. Electron. Mag.* **2016**, *10*, 25–39. [CrossRef]
10. Zhan, S.; Chong, A. Building Occupancy and Energy Consumption: Case Studies across Building Types. *Energy Built Environ.* 2020. [CrossRef]
11. Ramirez, A.D.; Boero, A.; Rivela, B.; Melendres, A.M.; Espinoza, S.; Salas, D.A. Life Cycle Methods to Analyze the Environmental Sustainability of Electricity Generation in Ecuador: Is Decarbonization the Right Path? *Renew. Sustain. Energy Rev.* **2020**, *134*, 110373. [CrossRef]
12. Hidalgo-León, R.; Litardo, J.; Urquizo, J.; Moreira, D.; Singh, P.; Soriano, G. Some Factors Involved in the Improvement of Building Energy Consumption: A Brief Review. In Proceedings of the 2019 IEEE Fourth Ecuador Technical Chapters Meeting (ETCM), Guayaquil, Ecuador, 11–15 November 2019; pp. 1–6.
13. DeForest, N.; Shehabi, A.; Selkowitz, S.; Milliron, D.J. A Comparative Energy Analysis of Three Electrochromic Glazing Technologies in Commercial and Residential Buildings. *Appl. Energy* **2017**, *192*, 95–109. [CrossRef]
14. Litardo, J.; Macías, J.; Hidalgo-León, R.; Cando, M.G.; Soriano, G. Measuring the Effect of Local Commercial Roofing Samples on the Thermal Behavior of Social Interest Dwelling Located in Different Climates in Ecuador. In Proceedings of the ASME 2019 International Mechanical Engineering Congress & Exhibition, Salt Lake City, UT, USA, 11–14 November 2019.
15. Doulos, L.T.; Kontadakis, A.; Madias, E.N.; Sinou, M.; Tsangrassoulis, A. Minimizing Energy Consumption for Artificial Lighting in a Typical Classroom of a Hellenic Public School Aiming for near Zero Energy Building Using LED DC Luminaires and Daylight Harvesting Systems. *Energy Build.* **2019**, *194*, 201–217. [CrossRef]
16. Wagiman, K.R.; Abdullah, M.N.; Hassan, M.Y.; Radzi, N.H.M.; Kwang, T.C. Lighting System Control Techniques in Commercial Buildings: Current Trends and Future Directions. *J. Build. Eng.* **2020**, 101342. [CrossRef]
17. Allouhi, A.; El Fouih, Y.; Kousksou, T.; Jamil, A.; Zeraouli, Y.; Mourad, Y. Energy Consumption and Efficiency in Buildings: Current Status and Future Trends. *J. Clean. Prod.* **2015**, *109*, 118–130. [CrossRef]
18. Macias, J.; Iturburu, L.; Rodriguez, C.; Agdas, D.; Boero, A.; Soriano, G. Embodied and Operational Energy Assessment of Different Construction Methods Employed on Social Interest Dwellings in Ecuador. *Energy Build.* **2017**, *151*, 107–120. [CrossRef]
19. Litardo, J.; Palme, M.; Borbor-Cordova, M.; Caiza, R.; Macias, J.; Hidalgo-Leon, R.; Soriano, G. Urban Heat Island Intensity and Buildings' Energy Needs in Duran, Ecuador: Simulation Studies and Proposal of Mitigation Strategies. *Sustain. Cities Soc.* **2020**, *62*, 102387. [CrossRef]
20. Franzetti, C.; Fraisse, G.; Achard, G. Influence of the Coupling between Daylight and Artificial Lighting on Thermal Loads in Office Buildings. *Energy Build.* **2004**, *36*, 117–126. [CrossRef]
21. Yahiaoui, A. Experimental Study on Modelling and Control of Lighting Components in a Test-Cell Building. *Sol. Energy* **2018**, *166*, 390–408. [CrossRef]
22. Han, H.J.; Mehmood, M.U.; Ahmed, R.; Kim, Y.; Dutton, S.; Lim, S.H.; Chun, W. An Advanced Lighting System Combining Solar and an Artificial Light Source for Constant Illumination and Energy Saving in Buildings. *Energy Build.* **2019**, *203*, 109404. [CrossRef]
23. Kruisselbrink, T.W.; Dangol, R.; van Loenen, E.J. A Comparative Study between Two Algorithms for Luminance-Based Lighting Control. *Energy Build.* **2020**, *228*, 110429. [CrossRef]
24. Delvaeye, R.; Ryckaert, W.; Stroobant, L.; Hanselaer, P.; Klein, R.; Breesch, H. Analysis of Energy Savings of Three Daylight Control Systems in a School Building by Means of Monitoring. *Energy Build.* **2016**, *127*, 969–979. [CrossRef]
25. Kunwar, N.; Cetin, K.S.; Passe, U.; Zhou, X.; Li, Y. Energy Savings and Daylighting Evaluation of Dynamic Venetian Blinds and Lighting through Full-Scale Experimental Testing. *Energy* **2020**, *197*, 117190. [CrossRef]
26. Somasundaram, S.; Chong, A.; Wei, Z.; Thangavelu, S.R. Energy Saving Potential of Low-e Coating Based Retrofit Double Glazing for Tropical Climate. *Energy Build.* **2020**, *206*, 109570. [CrossRef]

27. Marino, C.; Nucara, A.; Pietrafesa, M. Does Window-to-Wall Ratio Have a Significant Effect on the Energy Consumption of Buildings? A Parametric Analysis in Italian Climate Conditions. *J. Build. Eng.* **2017**, *13*, 169–183.
28. Wang, Y.; Wang, R.; Li, G.; Peng, C. An Investigation of Optimal Window-to-Wall Ratio Based on Changes in Building Orientations for Traditional Dwellings. *Sol. Energy* **2020**, *195*, 64–81.
29. Troup, L.; Phillips, R.; Eckelman, M.J.; Fannon, D. Effect of Window-to-Wall Ratio on Measured Energy Consumption in US Office Buildings. *Energy Build.* **2019**, *203*, 109434. [CrossRef]
30. Alghoul, S.K.; Rijabo, H.G.; Mashena, M.E. Energy Consumption in Buildings: A Correlation for the Influence of Window to Wall Ratio and Window Orientation in Tripoli, Libya. *J. Build. Eng.* **2017**, *11*, 82–86. [CrossRef]
31. Xue, P.; Li, Q.; Xie, J.; Zhao, M.; Liu, J. Optimization of Window-to-Wall Ratio with Sunshades in China Low Latitude Region Considering Daylighting and Energy Saving Requirements. *Appl. Energy* **2019**, *233*, 62–70. [CrossRef]
32. Praprost, M.; Fleming, K.A.; Dahlhausen, M. *ENERGY STAR for Tenants: An Online Energy Estimation Tool for Commercial Office Building Tenants*; National Renewable Energy Lab. (NREL): Golden, CO, USA, 2020.
33. Luddeni, G.; Krarti, M.; Pernigotto, G.; Gasparella, A. An Analysis Methodology for Large-Scale Deep Energy Retrofits of Existing Building Stocks: Case Study of the Italian Office Building. *Sustain. Cities Soc.* **2018**, *41*, 296–311. [CrossRef]
34. Kong, M.; Joo, H.; Kwak, H. Experimental Identification of Effects of Using Dual Airflow Path on the Performance of Roof-Type BAPV System. *Energy Build.* **2020**, *226*, 110403. [CrossRef]
35. Aly, A.M.; Chokwitthaya, C.; Poche, R. Retrofitting Building Roofs with Aerodynamic Features and Solar Panels to Reduce Hurricane Damage and Enhance Eco-Friendly Energy Production. *Sustain. Cities Soc.* **2017**, *35*, 581–593. [CrossRef]
36. Wang, Y.; Wang, D.; Liu, Y. Study on Comprehensive Energy-Saving of Shading and Photovoltaics of Roof Added PV Module. *Energy Procedia* **2017**, *132*, 598–603. [CrossRef]
37. Aparicio-Gonzalez, E.; Domingo-Irigoyen, S.; Sánchez-Ostiz, A. Rooftop Extension as a Solution to Reach NZEB in Building Renovation. Application through Typology Classification at a Neighborhood Level. *Sustain. Cities Soc.* **2020**, *57*, 102109. [CrossRef]
38. Magrini, A.; Lentini, G.; Cuman, S.; Bodrato, A.; Marenco, L. From Nearly Zero Energy Buildings (NZEB) to Positive Energy Buildings (PEB): The next Challenge-The Most Recent European Trends with Some Notes on the Energy Analysis of a Forerunner PEB Example. *Dev. Built Environ.* **2020**, *3*, 100019. [CrossRef]
39. Asdrubali, F.; Baggio, P.; Prada, A.; Grazieschi, G.; Guattari, C. Dynamic Life Cycle Assessment Modelling of a NZEB Building. *Energy* **2020**, *191*, 116489. [CrossRef]
40. Feng, W.; Zhang, Q.; Ji, H.; Wang, R.; Zhou, N.; Ye, Q.; Hao, B.; Li, Y.; Luo, D.; Lau, S.S.Y. A Review of Net Zero Energy Buildings in Hot and Humid Climates: Experience Learned from 34 Case Study Buildings. *Renew. Sustain. Energy Rev.* **2019**, *114*, 109303. [CrossRef]
41. Sudhakar, K.; Winderl, M.; Priya, S.S. Net-Zero Building Designs in Hot and Humid Climates: A State-of-Art. *Case Stud. Therm. Eng.* **2019**, *13*, 100400. [CrossRef]
42. Ghenai, C.; Bettayeb, M. Modelling and Performance Analysis of a Stand-Alone Hybrid Solar PV/Fuel Cell/Diesel Generator Power System for University Building. *Energy* **2019**, *171*, 180–189. [CrossRef]
43. Liu, J.; Chen, X.; Cao, S.; Yang, H. Overview on Hybrid Solar Photovoltaic-Electrical Energy Storage Technologies for Power Supply to Buildings. *Energy Convers. Manag.* **2019**, *187*, 103–121. [CrossRef]
44. Peel, M.C.; Finlayson, B.L.; McMahon, T.A. Updated World Map of the Köppen-Geiger Climate Classification. In *Hydrology and Earth System Sciences Discussions*; European Geosciences Union: Munich, Germany, 2007; Volume 4, pp. 439–473.
45. Guevara, G.; Soriano, G.; Mino-Rodriguez, I. Thermal Comfort in University Classrooms: An Experimental Study in the Tropics. *Build. Environ.* **2020**, *187*, 107430. [CrossRef]
46. INEC. *Proyección de La Población Ecuatoriana, Por Años Calendario, Según Cantones 2010–2020*; INEC: Quito, Ecuador, 2020.
47. Meteotest. *Meteonorm 7.3*; Meteotest: Bern, Switzerland, 2018.
48. Energy Design Tools UCLA. *Climate Consultant 6.0 Sofware*; UCLA: Los Ángeles, CA, USA, 2020.
49. ESPOL. Escuela Superior Politécnica Del Litoral (ESPOL). Available online: http://www.espol.edu.ec/ (accessed on 22 December 2020).
50. Agencia de Regulación y Control de Electricidad (ARCONEL). *Pliego Tarifario Para Las Empresas Eléctricas de Distribución Codificado*; ARCONEL: Quito, Ecuador, 2019.
51. Fleming, K.; Long, N.; Swindler, A. *Building Component Library: An Online Repository to Facilitate Building Energy Model Creation*; National Renewable Energy Lab. (NREL): Golden, CO, USA, 2012.
52. ANSI; ASHRAE; IES. *Standard 90.1-2019 Energy Standard for Buildings Except Low-Rise Residential Buildings*; ASHRAE: Atlanta, Georgia, USA, 2019.
53. Kim, D.; Cho, H.; Koh, J.; Im, P. Net-Zero Energy Building Design and Life-Cycle Cost Analysis with Air-Source Variable Refrigerant Flow and Distributed Photovoltaic Systems. *Renew. Sustain. Energy Rev.* **2020**, *118*, 109508. [CrossRef]
54. Homer Energy. *Homer PRO*; Homer Energy: Boulder, CO, USA, 2020.
55. Maxeon Solar Technologies. *Sun Power Maxeon 2—360W*; Maxeon Solar Technologies: Singapore, 2020.
56. ProViento, Paneles Solares. Available online: https://proviento.com.ec/10-paneles-solares (accessed on 22 December 2020).
57. Jamil, W.J.; Rahman, H.A.; Shaari, S.; Desa, M.K.M. Modeling of Soiling Derating Factor in Determining Photovoltaic Outputs. *IEEE J. Photovolt.* **2020**, *10*, 1417–1423. [CrossRef]

58. Vallejo, C.; Villacreses, G.; Vásquez, F.; Godoy, F. *Evaluación Comparativa de Los Consumos Energéticos de Edificaciones Públicas En La Región Costa y Galápagos*; Instituto Nacional de Eficiencia Energértica y Energías Renovables (INER): Quito, Ecuador, 2018.
59. Litardo, J.; Hidalgo-León, R.; Macías, J.; Delgado, K.; Soriano, G. *Estimating Energy Consumption and Conservation Measures for ESPOL Campus Main Building Model Using EnergyPlus*; IEEE CONCAPAN: Ciudad de Guatemala, Guatemala, 2019; pp. 1–6.
60. Cámara de la Construcción de Guayaquil. Available online: http://www.cconstruccion.net/precios.html (accessed on 22 December 2020).
61. Compras Públicas Ecuador, Sistema Oficial de Contratación Pública. Available online: https://www.compraspublicas.gob.ec/ProcesoContratacion/compras/ (accessed on 22 December 2020).
62. Syahputra, R.; Soesanti, I. Planning of Hybrid Micro-Hydro and Solar Photovoltaic Systems for Rural Areas of Central Java, Indonesia. *J. Electr. Comput. Eng.* **2020**, *2020*, 5972342. [CrossRef]
63. Hee, W.J.; Alghoul, M.A.; Bakhtyar, B.; Elayeb, O.; Shameri, M.A.; Alrubaih, M.S.; Sopian, K. The Role of Window Glazing on Daylighting and Energy Saving in Buildings. *Renew. Sustain. Energy Rev.* **2015**, *42*, 323–343. [CrossRef]
64. Shishegar, N.; Boubekri, M. Quantifying Electrical Energy Savings in Offices through Installing Daylight Responsive Control Systems in Hot Climates. *Energy Build.* **2017**, *153*, 87–98. [CrossRef]
65. Whang, A.J.-W.; Yang, T.-H.; Deng, Z.-H.; Chen, Y.-Y.; Tseng, W.-C.; Chou, C.-H. A Review of Daylighting System: For Prototype Systems Performance and Development. *Energies* **2019**, *12*, 2863. [CrossRef]
66. Köster, H. Daylighting Controls, Performance, and Global Impacts. *Sustain. Built Environ.* **2020**, 383–429. [CrossRef]
67. Kim, H.G.; Kim, H.J.; Jeon, C.H.; Chae, M.W.; Cho, Y.H.; Kim, S.S. Analysis of Energy Saving Effect and Cost Efficiency of ECMs to Upgrade the Building Energy Code. *Energies* **2020**, *13*, 4955. [CrossRef]
68. Papadopoulos, S.; Kontokosta, C.E.; Vlachokostas, A.; Azar, E. Rethinking HVAC Temperature Setpoints in Commercial Buildings: The Potential for Zero-Cost Energy Savings and Comfort Improvement in Different Climates. *Build. Environ.* **2019**, *155*, 350–359. [CrossRef]
69. Bienvenido-Huertas, D.; Sánchez-García, D.; Pérez-Fargallo, A.; Rubio-Bellido, C. Optimization of Energy Saving with Adaptive Setpoint Temperatures by Calculating the Prevailing Mean Outdoor Air Temperature. *Build. Environ.* **2020**, *170*, 106612. [CrossRef]
70. Ge, J.; Wu, J.; Chen, S.; Wu, J. Energy Efficiency Optimization Strategies for University Research Buildings with Hot Summer and Cold Winter Climate of China Based on the Adaptive Thermal Comfort. *J. Build. Eng.* **2018**, *18*, 321–330. [CrossRef]
71. Al-Masrani, S.M.; Al-Obaidi, K.M.; Zalin, N.A.; Isma, M.A. Design Optimisation of Solar Shading Systems for Tropical Office Buildings: Challenges and Future Trends. *Sol. Energy* **2018**, *170*, 849–872. [CrossRef]
72. Alhuwayil, W.K.; Mujeebu, M.A.; Algarny, A.M.M. Impact of External Shading Strategy on Energy Performance of Multi-Story Hotel Building in Hot-Humid Climate. *Energy* **2019**, *169*, 1166–1174. [CrossRef]
73. Al Touma, A.; Ouahrani, D. Shading and Day-Lighting Controls Energy Savings in Offices with Fully-Glazed Façades in Hot Climates. *Energy Build.* **2017**, *151*, 263–274. [CrossRef]
74. Wati, E.; Meukam, P.; Nematchoua, M.K. Influence of External Shading on Optimum Insulation Thickness of Building Walls in a Tropical Region. *Appl. Therm. Eng.* **2015**, *90*, 754–762. [CrossRef]
75. ARCONEL. *Resolucion Nro. ARCONEL-042/18*; Agencia de Regulacion y Control de Electricidad: Quito, Ecuador, 2018.
76. Vaca-Revelo, D.; Ordóñez, F. *Mapa Solar del Ecuador 2019*; Escuela Politécnica Nacional (EPN): Quito, Ecaudor, 2019.
77. Korsavi, S.S.; Zomorodian, Z.S.; Tahsildoost, M. Energy and Economic Performance of Rooftop PV Panels in the Hot and Dry Climate of Iran. *J. Clean. Prod.* **2018**, *174*, 1204–1214. [CrossRef]
78. Eisenhower, B.; O'Neill, Z.; Fonoberov, V.A.; Mezić, I. Uncertainty and Sensitivity Decomposition of Building Energy Models. *J. Build. Perform. Simul.* **2012**, *5*, 171–184. [CrossRef]
79. Fumo, N.; Mago, P.; Luck, R. Methodology to Estimate Building Energy Consumption Using EnergyPlus Benchmark Models. *Energy Build.* **2010**, *42*, 2331–2337. [CrossRef]
80. ASHRAE. *Guideline 14-2002, Measurement of Energy and Demand Savings*; Technical Report; ASHRAE: Atlanta, GA, USA, 2002.
81. Judkoff, R.; Neymark, J. *Model Validation and Testing: The Methodological Foundation of ASHRAE Standard 140*; National Renewable Energy Lab. (NREL): Golden, CO, USA, 2006.

Article

Non-Intrusive Load Disaggregation Based on a Multi-Scale Attention Residual Network

Liguo Weng [1], Xiaodong Zhang [1], Junhao Qian [1], Min Xia [1,*], Yiqing Xu [2] and Ke Wang [3]

1 Jiangsu Collaborative Innovation Center on Atmospheric Environment and Equipment Technology, Nanjing University of Information Science and Technology, Nanjing 210044, China; 002311@nuist.edu.cn (L.W.); 20181223103@nuist.edu.cn (X.Z.); 20181223052@nuist.edu.cn (J.Q.)
2 College of Information Science and Technology, Nanjing Forestry University, Nanjing 210037, China; yiqingxu@njfu.edu.cn
3 China Electric Power Research Institute, Nanjing 210003, China; wangke@epri.sgcc.com.cn
* Correspondence: xiamin@nuist.edu.cn

Received: 30 November 2020; Accepted: 17 December 2020; Published: 21 December 2020

Abstract: Non-intrusive load disaggregation (NILD) is of great significance to the development of smart grids. Current energy disaggregation methods extract features from sequences, and this process easily leads to a loss of load features and difficulties in detecting, resulting in a low recognition rate of low-use electrical appliances. To solve this problem, a non-intrusive sequential energy disaggregation method based on a multi-scale attention residual network is proposed. Multi-scale convolutions are used to learn features, and the attention mechanism is used to enhance the learning ability of load features. The residual learning further improves the performance of the algorithm, avoids network degradation, and improves the precision of load decomposition. The experimental results on two benchmark datasets show that the proposed algorithm has more advantages than the existing algorithms in terms of load disaggregation accuracy and judgments of the on/off state, and the attention mechanism can further improve the disaggregation accuracy of low-frequency electrical appliances.

Keywords: load disaggregation; multi-scale; attention mechanism; residual network

1. Introduction

Load disaggregation technology is a key technology in smart grids [1]. Traditional load monitoring adopts intrusive methods, which are able to obtain accurate and reliable data with low data noise [2], but they are difficult to be accepted by users due to their high implementation costs. Non-intrusive methods can provide detailed information for residents in time, and have the advantages of low cost and easy implementation. According to this technology, the power consumption behaviors of users can be analyzed, and users can be guided toward a reasonable consumption of electricity and hence reduce their power consumption costs. With the continuous development of power demand side management [3], big data analysis, and other technologies, non-intrusive load disaggregation is attracting more attention.

The microgrid is an important manifestations of the smart grid. With the development of clean energy, such as solar and wind energy, and energy internet technology, the microgrid has emerged. It is a small power system with distributed power sources, which can realize a highly reliable supply of multiple energy sources and improve the quality of the power supply [4]. As NILM technology becomes more mature, the intelligent dispatching of the microgrid can be realized through automation in the future to improve the effective utilization of power resources, ensure the stable economic operation of a power system, and avoid the unnecessary waste of power resources. Therefore, NILM technology is important.

The concept of non-intrusive load monitoring was firstly proposed by Hart [5]. It mainly uses non-intrusive load disaggregation (NILD). In this method, the total power consumption is disaggregated to each individual electrical appliance. Hart proposed the concept of "load characteristics", which he defined as the information change of the electrical power of an appliance in operation. Hart further used the steady-state load characteristics [6] to design a simple NILD system to decompose power. However, effective features extracted by the algorithm were limited, and large disaggregation errors occurred easily.

At present, combinatorial optimization (CO) methods and pattern recognition algorithms are the main algorithms for realizing non-intrusive load disaggregation. Among them, NILD based on a combinatorial optimization algorithm [7] determines the power consumption value of each appliance by investigating load characteristics as well as error comparisons between power states of combined appliances and the total power. Chang [8] and Lin [9] used the Particle Swarm Optimization (PSO) algorithm to solve the disaggregation problem based on the steady state current on a few electrical appliances, but the disaggregation result error was large. In order to solve the NILD problem, Piga [10] proposed a sparse optimization method to improve the disaggregation accuracy. The combinatorial optimization method is essentially an NP-hard problem, so its efficiency is a challenge. In addition, the optimization theory could be only used to analyze discrete states of electrical appliances, so it is difficult to model loads with large load fluctuations.

With the development of machine learning, pattern recognition algorithms have been applied to NILD. Batra [11] solved the depolymerization problem of low-power appliances using K-nearest neighbor regression (KNN) [12], but the algorithm could not solve the problem of the large power difference between appliances. Kolter [13] used the sparse coding algorithm to learn the power consumption models of each electrical appliance and used these models to predict the power of each electrical appliance. Johnson [14] used unsupervised learning for NILD, and this model had a high training speed. However, compared to the supervised algorithms, the ability of Johnson's method's to identify complex continuous state loads was limited because of the lack of prior knowledge. Kim [15] used the multi-factor hidden Markov algorithm to disaggregate the continuous value of each electrical appliance according to the given total power data. Some excellent machine learning algorithms, such as the support vector machine [16] and the adaboost algorithm [17], achieved certain processes, but these methods shared the same problem: a large number of load characteristics were required for identification, a requirement that was often difficult to meet in practice. Different from traditional methods, the deep learning method [18,19] is able to automatically extract features from original data [20]. In Kelly's [21] experiment, various NILD algorithms using deep learning were proposed, such as the Delousing AutoEncoder (DAE), the long-short term memory network (LSTM), the gatedrecurrent unit (GRU), the Factorial Hidden Markov model (FHMM), and the CO method. The DAE algorithm was proven to have good disaggregation results. Zhang [22] used two convolutional neural network algorithms for load disaggregation. Compared with Kelly's method, the two CNN methods, sequence-to-sequence and sequence-to-point, achieved better performance [23], but their layer numbers were small, and hence there were unable to extract higher level load characteristics. In the above methods, the CO algorithm, the DAE, and the two CNN methods were all trained by low-frequency data from the REDD dataset, which was first processed by the NILM-TK toolkit. The sampling interval of the data was 3 s. With an improvement of the model structure, Yang [24] proposed a semisupervised deep learning framework based on BiLSTM and the temporal convolutional network for multi-label load classification. Akhilesh [25] proposed a multilayer deep neural network based on the sequence-to-sequence methodology, and the algorithm, by reading the daily load profile for the total power consumption, could identify the state of the appliances according to the device-specific power signature. Since the neural networks were only trained for each appliance and the computational cost was high, Anthony [26] proposed UNet-NILM for multi-task appliances' state detection and power estimation, which had a good performance compared with traditional single-task deep learning.

The innovation of the algorithm proposed in this paper lies in the following: The multi-scale structure is used to extract different load information according to the characteristics of load disaggregation.

The attention mechanism is used to fuse the load information at different scales to further enhance the feature extraction ability of the network, especially for the extraction of electrical features that are not frequently used. The overall architecture uses a skip connection of the residual network [27] to improve network performance. Experimental results on two benchmark datasets show that our method is superior to other present methods.

2. Multi-Scale Attention Residual Network

2.1. Deep Residual Network

The depth of the neural network [28] is crucial to the performance of the model. Increasing the depth of the neural network is helpful to extract more complex features and improve the accuracy of the algorithm. However, at the same time, when the network reaches a certain depth, further deepening would make the training accuracy saturated or even reduced. Traditional methods dealing with gradient disappearance and gradient explosion, e.g., the activation function Relu and batch normalization, are able to alleviate these problems to a certain extent, but not fundamentally solve the problems.

The deep residual network (Resnet) [29] uses the idea of cross-layer connection. If those behind the deep network are identity mapping, the model would degenerate directly into a shallow network, and it is difficult to use the neural network stacked with hidden layers to fit into a potential identity mapping function $H(x) = x$. However, if the network is designed to be $H(x) = F(x) + x$, the procedure can be translated into learning a residual function $F(x) = H(x) - x$. When $F(x) = 0$, it presents an identity map $F(x) = H(x)$. The residual network is a structure that outputs the features of front layers to back layers. By introducing this structure, a neural network is able to perform well.

Assuming that the input of a forward neural network is x, the dimension is $H \times W \times C$, and the expected output of the network is $H(x)$. The residual structure $F(x) + x$ can be realized by a forward neural network and a skip connection. When the number of input and output channels is the same, the dimension of $H(x)$ is also $H \times W \times C$. A skip connection means that one or more layers of the network are bypassed. A skip connection performs only one identity mapping and adds its outputs to the outputs of stack layers.

The residual structure is shown in Figure 1. Each structure has two layers, where W_1 and W_2 are the weight matrixes. The input is sequentially multiplied by the W_1 matrix of the first layer, activated by the relu function, and then multiplied by the W_2 matrix of the second layer to obtain the forward output. The forward neural network of the residual structure could be expressed as

$$F(x) = W_2\sigma(W_1 x), \tag{1}$$

where σ is the activation function. The right hand is the skip connection x of the residual structure, and the final output is obtained after a summation operation. The formula is as follows:

$$y = F(x, W_i) + x, \tag{2}$$

where $F(x, W_i)$ is the residual mapping function that needs to be studied. W_i represents the weight matrix of the hidden layer. When the input and output dimensions need to be changed, a linear transformation $W \cdot x$ could be performed through the input at the skip connection residual structure. Thus, the same figure of input and output characteristics could be expressed as

$$H(x) = F(x, W_i) + W_s x. \tag{3}$$

According to Equation (3), it could be noted that for a deeper layer L, its relationship with the l layer could be expressed as

$$x_L = x_l + \sum_{i-1}^{L-1} F(x_l, W_l), \tag{4}$$

where layer x_L and layer x_l are the residual unit inputs of layer L and layer l. According to Equation (4), inputs of the residual unit in layer L could be expressed as the sum of inputs of a shallow residual unit and all the complex mappings. The calculation power needed of the sum is much less than that of the quadratics.

With a loss function, and according to the chain rule of back propagation, we can obtain

$$\frac{\partial \varepsilon}{\partial x_l} = \frac{\partial \varepsilon}{\partial x_L}\frac{\partial x_L}{\partial x_l} = \frac{\partial \varepsilon}{\partial x_L}(1 + \frac{\partial}{\partial x_l}\sum_{L-1}^{i=1} F(x_i, W_l)). \tag{5}$$

This means that continuous multiplications generated in the network are replaced by plus operations, and the problem of gradient disappearance is well solved. The use of residual structure is able to increase the network depth and extract deeper load characteristics from the data, thus improving the accuracy of disaggregation algorithms. Based on the residual network, in view of the characteristics of load disaggregation, we replaced the convolutional layer with the multi-scale structure and the attention structure. Therefore, we proposed the MSA-Resnet.

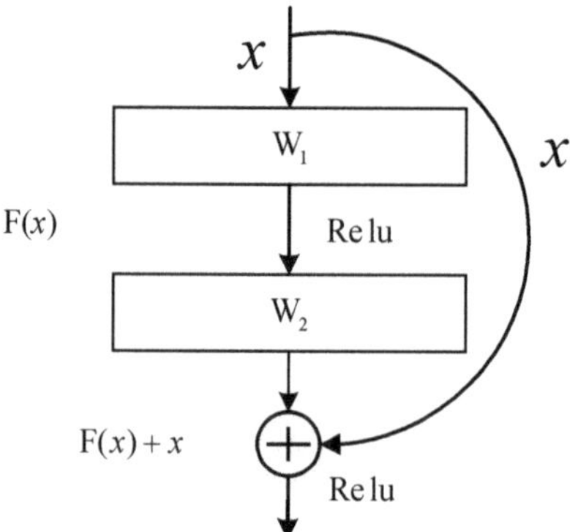

Figure 1. The residual structure.

2.2. Attention Mechanism

A convolution kernel, the core of a CNN, is generally regarded as aggregating information spatially and channel-wise on local receptive fields. A CNN is composed of a series of convolution layers, non-linear layers, and down-sampling layers, among others, and captures required features from global receptive fields [30].

However, in order to obtain better network performance, a squeeze and excitation attention mechanism is used in the network [31]. Its structure is shown in Figure 2. One of the novelties of the algorithm in this paper lies in the use of the attention mechanism in the residual structure to further improve the feature extraction ability of the network, especially for the extraction of electrical features that are not frequently used. This attention mechanism is different from the previous structure, as it improves performance from feature channels. The first 3D matrix in Figure 2 is composed of a feature graph C with a size of $H \times W$.

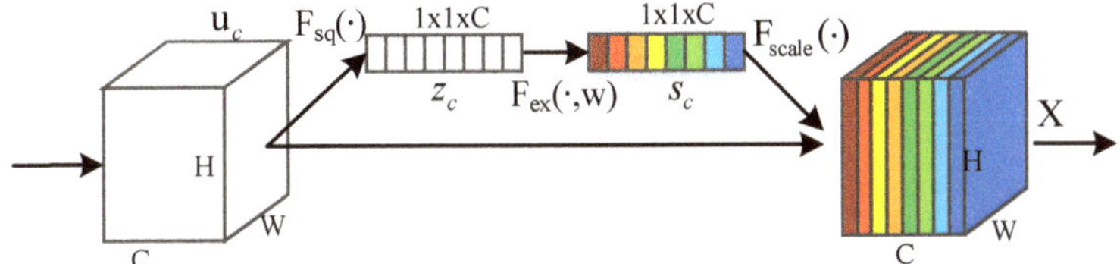

Figure 2. The attention block.

According to Figure 2, the spliced feature map of a multi-scale module is obtained through the global pooling layer to obtain attention vector z_c, which is compressed from the spatial dimension to obtain the global receptive field. z_c is a high-dimensional vector containing the low-order global feature information of the feature map, and its dimension is $1 \times 1 \times C$. The expression equation is as follows:

$$z_c = F_{sq}(u_c) = \frac{1}{h \times w}(\sum_{h}^{i=1}\sum_{w}^{j=1} u_c(i,j)). \tag{6}$$

Next, two fully connected layers are applied to establish correlations between channels as excitation and to output the same number of weights as the input feature. The equation is

$$s_c = F_{ex}(z_c, w_2) = \delta(w_2\sigma(w_1 z_c), \tag{7}$$

where the dimensions of w_1 and w_2 are all $\frac{C}{s} \times C$, s represents the scaling coefficient, and the attention vector s_c is obtained after being activated by the Relu function and the Sigmoid function. It is a high-dimensional vector of high-order global feature information obtained on the basis of attention vector z_c. The attention vector s_c further represents the change of load feature information in the channel dimension, and its dimension is also $1 \times 1 \times C$. A Sigmoid activation function is similar to a gating processing mechanism, which generates different weights for each feature channel of the attention vector s_c. Finally, the original three-dimensional matrix u_c is multiplied to complete the weight recalibration. Therefore, the importance of each load feature is obtained according to the feature map [32]. Finally, the useful load characteristics are enhanced and the less useful load characteristics are suppressed, which can improve the accuracy of load disaggregation of low-usage appliances. The equation is

$$X = F_{scale}(u_c, s_c) = u_c \cdot s_c. \tag{8}$$

2.3. Multi-Scale Attention Resnet Based NILD

From the point of view of neural networks, the problem of NILM can be interpreted in this way. Assuming that $Y(t)$ is the sum of all the active power consumption of appliances in the household, it can be expressed as the following formula:

$$Y(t) = \sum_{i}^{I} X_i(t) + e(t). \tag{9}$$

In the formula, $X_i(t)$ represents the power data of electrical equipment i at time t, I represents the number of electrical equipment, and e represents the model noise. Therefore, there is a pair of data (X, Y), a model can be trained to represent the relationship between X and Y. $X = f(Y)$ is a non-linear regression problem, and the neural network is an effective method to learn the function f.

The overall network structure of the multi-scale attention residual network (MSA-Resnet) proposed in this paper is shown in Figure 3.

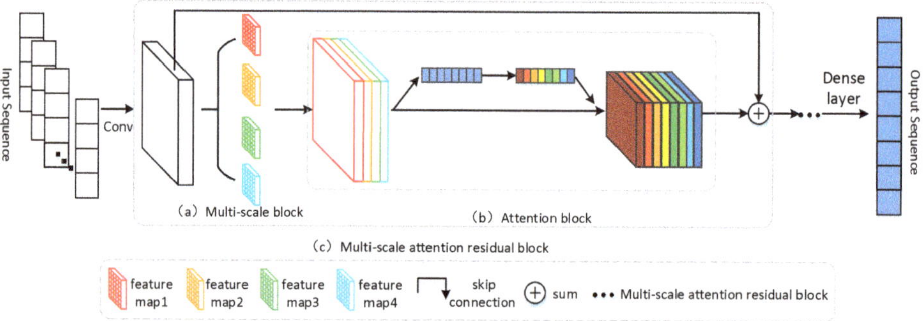

Figure 3. The MSA-Resnet overall structure.

(a) The multi-scale module is composed of convolution kernels with sizes of 3×1, 5×1, and 1×1 and the pooling layer [33]. By combining (b) the attention block and the residual structure, (c) the multi-scale attention residual block is formed. The structure of (a) is shown in Figure 4.

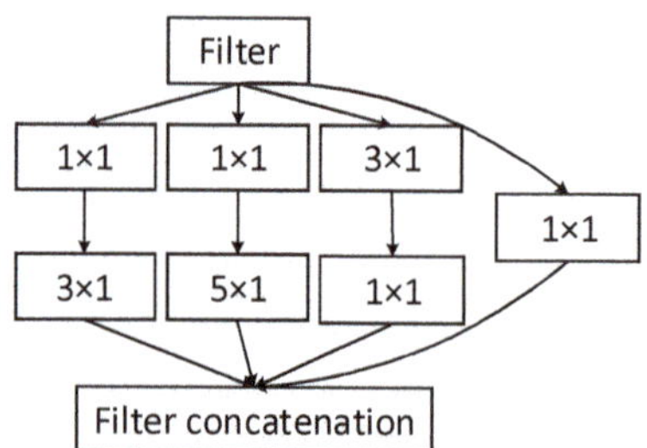

Figure 4. Multi-scale block.

All convolution cores of original residual elements are 3×1 in size, which makes it impossible for convolution layers to observe load data from multiple scales, and difficult to obtain more abundant load features. The multi-scale module first goes though a 3×1 convolution, followed by four branches. The first branch uses a 1×1 convolution to increase the load characteristics transformation [34], and a 3×1 convolution is then applied to obtain a feature map (map1). The second branch is convolved at 1×1, and a 5×1 convolution is then added to obtain map2. The third branch is pooled at 3×1 to obtain map3. The fourth branch uses a 1×1 convolution to obtain map4. Finally, these feature maps are concatenated to input vectors for the attention module. Because the actual load power has a large number of different gear positions, switch starts and stops, and operating characteristics, the multi-scale feature method can improve the network's ability to extract load characteristics and increase the diversity of different scales of the network, thus improving the accuracy of non-intrusive load disaggregation.

In the network structure of MSA-Resnet, nine multi-scale attention residual blocks are used. Forty convolution cores are used in the first three blocks, 60 convolution cores are used in the forth to sixth blocks, and 80 convolution cores are allocated in the last three. The first convolution layer and each output part of multi-scale attention residual block is activated by an activation function Leaky-Relu. Relu and Leaky-Relu [35] are shown in Figure 5:

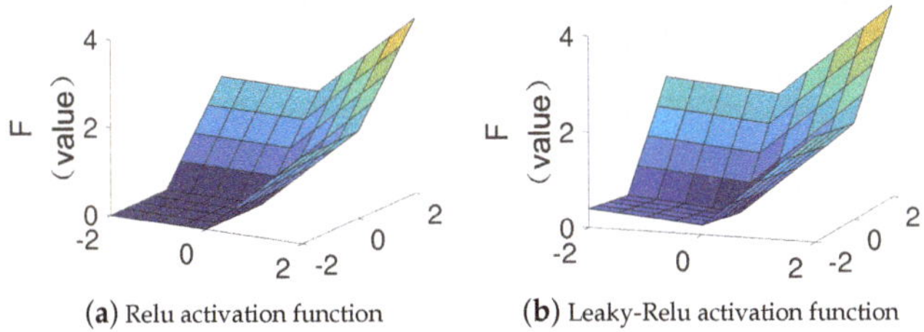

(**a**) Relu activation function (**b**) Leaky-Relu activation function

Figure 5. The Comparison of Relu and Leaky-Relu.

The Relu activation function represents the "modified linear element," which could accelerate the convergence speed of the network. Its equation is

$$f(x) = max(0, x). \tag{10}$$

When the input is positive, the derivative is not zero, so the learning is based on the gradient. However, when the input is negative, the learning speed of Relu slows down and can even inactivate the neurons, such that it cannot follow new weights. Equation (11) represents the Leaky-Relu activation function, where $\lambda \in (0, 1)$ modifies the data distribution and retains the value of the negative axis. As a result, the information retention ability is improved without losing more load characteristics, and the gradient is guaranteed not to appear.

$$f(x) = \begin{cases} x & \text{if } x > 0 \\ \lambda x & \text{if } x < 0 \end{cases}. \tag{11}$$

3. Data Selection and Preprocessing

3.1. Data Sources

The experimental data in this paper is from the public dataset UK-DALE [36] and WikiEnergy [37]. The UK-DALE dataset is a public access dataset from the UK, and the sampling frequency is 1/6 Hz. The WikiEnergy dataset was produced by the Pecan Street company in the UK and contains the data of nearly 600 households. It includes the total power consumed by each household over a period of time and the power consumed by each individual electrical appliance. The sampling frequency of the dataset is 1/60 Hz [38]. Kettles, air conditioners, fridges, microwaves, washing machines, and dishwashers were chosen as non-intrusive load disaggregation tasks for the following reasons: (1) The power consumption of these electrical appliances is a large proportion of the total power consumption. They are representatives of electrical appliances. (2) Electrical appliances with low frequency and minimal power consumption used in the data are easily disturbed by noise and not easily disaggregated. (3) The power consumption of these six electrical devices includes mode disaggregation from simple to complex.

3.2. Data Preprocessing

Firstly, the NILM-TK toolkit was used to export the power data of the selected home appliances from the WikiEnergy database and UK-DALE database. We then created aggregate power profiles and used them as the experimental data. Secondly, different evaluation indexes often have different dimensions, and their values are quite wide ranged, and this may affect the analysis results. In order to eliminate the influence of these differences, data standardization is needed. Here, the maximum and minimum value normalization method was used to normalize the result between $[0, 1]$. The normalization equation is

$$x_t^* = \frac{x_t - X_{min}}{X_{max} - X_{min}}, \tag{12}$$

where x_t is the total unstandardized power at time t, X_{max} is the maximum of the total power sequence, X_{min} is the minimum of the total power sequence, and x_t^* is the standardized result at time t.

3.3. Sliding Window

Deep learning training relies on a large amount of data. The NILM-TK toolkit was used to process the database. We selected the desired electrical data and sorted it into an Excel file. The top 80% of the processed data was taken as training data. The total power sequence X was taken as the input sequence, and the individual electrical appliance Y was taken as the target sequence. The remaining 20% of the processed data was taken as testing data. In order to increase the training data of the network and improve the expression ability of the data, the data was processed by using a sliding window [39].

As shown in Figure 6a, the overlap sliding window [40] was used to process the total power sequence and the target sequence in the training data to increase the data samples. Assuming that the length of power sequence is M, a window of length N was cut from the original data with a sliding step of 1, and the sliding operation was then carried out to obtain $M - N + 1$ samples. However, as shown in Figure 6b, the non-overlap sliding window was used to process the testing data to save time. Assuming that the sequence length is H, $\frac{H}{N}$ samples could be obtained.

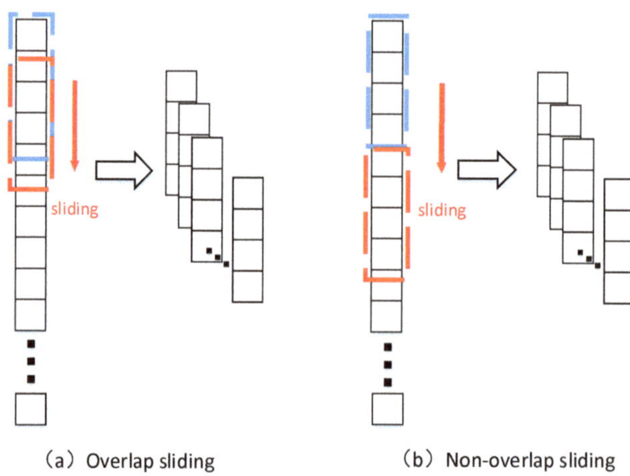

Figure 6. Sliding operation.

4. Result

This experiment used the Keras neural network framework. The computer processor was AMD2600, and the graphics card was 1060 6G. After data was standardized, the length of the sliding window was set to 200, the learning rate of the network was set to 0.001, and the Adam optimizer was selected as the network optimizer.

Kelly's experiments indicate that the DAE algorithm performs well in NILD, and Zhang C's work also shows a good performance of CNNs in sequence-to-sequence and sequence-to-point load disaggregation. From the WikiEnergy data, we selected the air conditioner, fridge, microwave, washing machine, and dishwasher from Household 25. From the UK-DALE dataset, the kettle, fridge, microwave, washing machine, and dishwasher of Household 5 were selected. In order to verify the effectiveness and stability of the algorithm proposed in this paper, four approaches were compared with the MSA-Resnet: the KNN, the DAE, the CNN sequence-to-sequence learning (CNN s-s), and the CNN sequence-to-point learning (CNN s-p). Firstly, the WikiEnergy dataset was tested. Figure 7 shows the disaggregation effect diagrams of five appliances of WikiEnergy from Household 25, and the actual power consumption data of these appliances. The figure compares the four disaggregation methods with the MSA-Resnet proposed in this paper.

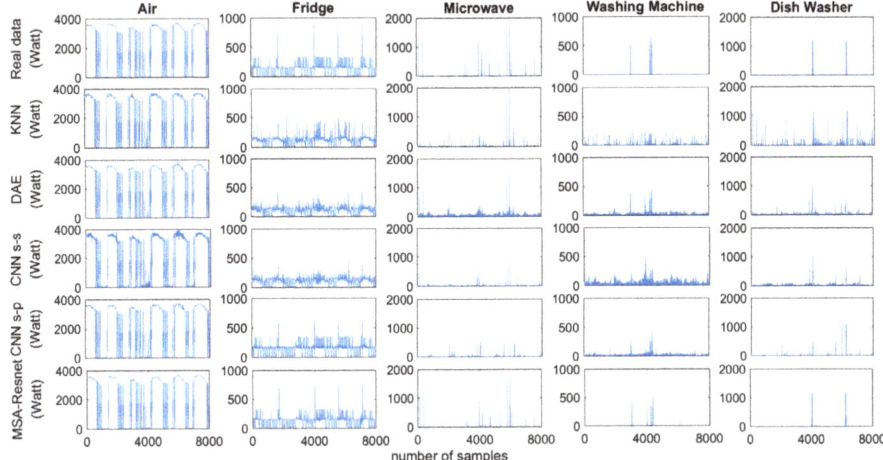

Figure 7. Comparison load disaggregation results of Household 25 in the WikiEnergy dataset.

In order to verify the effectiveness of the proposed method, two evaluation indexes were selected to evaluate the performance of the algorithm: the Mean Absolute Error (MAE) and the Signal Aggregate Error (SAE). The MAE evaluation index was used to measure the average error of power consumption and the actual power consumption of individual electrical appliances disaggregated at each moment. The MAE is expressed as the following:

$$MAE = \frac{1}{T}\sum_{T}^{t=1}|p_t - g_t|, \tag{13}$$

where g_t represents the actual power consumed by an appliance at time t, p_t represents the disaggregation power of the appliance at time t, and T represents the number of time points.

Equation (14) is the expression of the SAE, where $\hat{e}$ and e represent the power consumption predicted by disaggregation within a period of time and the real power consumption within a period of time. This index is helpful for daily electricity reports.

$$SAE = \frac{|\hat{e} - e|}{e}. \tag{14}$$

Figure 7 describes disaggregations of Household 25 in the WikiEnergy dataset. It can be seen that the above algorithms can basically achieve effective load disaggregation for the air conditioner. In the load disaggregation diagram of the fridge, the DAE and CNN s-s algorithms fluctuate greatly in the mean area of the appliance, compared with other algorithms. The KNN algorithm has the worst load disaggregation effect on the last three kinds of electrical appliances, so it could not realize an effective disaggregation of mutation points. For these three low-frequency electrical appliances, the load disaggregation of CNN s-s and CNN s-p algorithms are stable compared with the other two algorithms, but the load disaggregation of the CNN s-p method fluctuates greatly in the region of low power consumption. In summary, compared with other methods in load disaggregation, the MSA-Resnet shows the best performance on each electrical appliance, based on the power consumption curve.

Table 1 shows comparisons of MAE and SAE indexes of Household 25 load disaggregation in the WikiEnergy dataset. It can be seen that MSA-Resnet has obvious advantages in the disaggregations of the air conditioner, fridge, microwave, washing machine, and dishwasher. According to the MAE index, the MSA-Resnet performs better than the other four methods. For the SAE, the MSA-Resnet achieves the lowest value on the fridge, washing machine, and dishwasher, and accurate disaggregation of energy is achieved over a period of time. Combined with Figure 7 and Table 1, it can be inferred that the shallow CNN s-s and CNN s-p have difficulty accurately disaggregating the total power into the appliances with lower frequency. Compared with KNN and MSA-Resnet, the disaggregation errors of CNN s-s and CNN s-p are larger, because the structure of shallow CNNs is not able to extract deeper and more effective load characteristics, and their disaggregation effect is not as good as that of MSA-Resnet. There are two reasons for this: firstly, the residual is used to deepen the network and better enhance the ability to learn unbalanced samples; secondly, the ability to deal with low frequency appliances by multi-scale convolutions is strong. As can be seen in Figure 7, the overall disaggregation effect of the KNN on the washing machine is not good, but the disaggregation error is small in terms of two indicators. To explain this phenomenon, certain interval periods are selected for comparative analysis, as shown in Figure 8, the disaggregation comparison diagram shows each algorithm on each electrical appliance with a finer scale. The figure reflects the ability of the KNN to detect peak values. It can be seen in Figure 8b,c that the KNN is not able to accurately disaggregate mutation points, but it could process regions with a power close to 0.

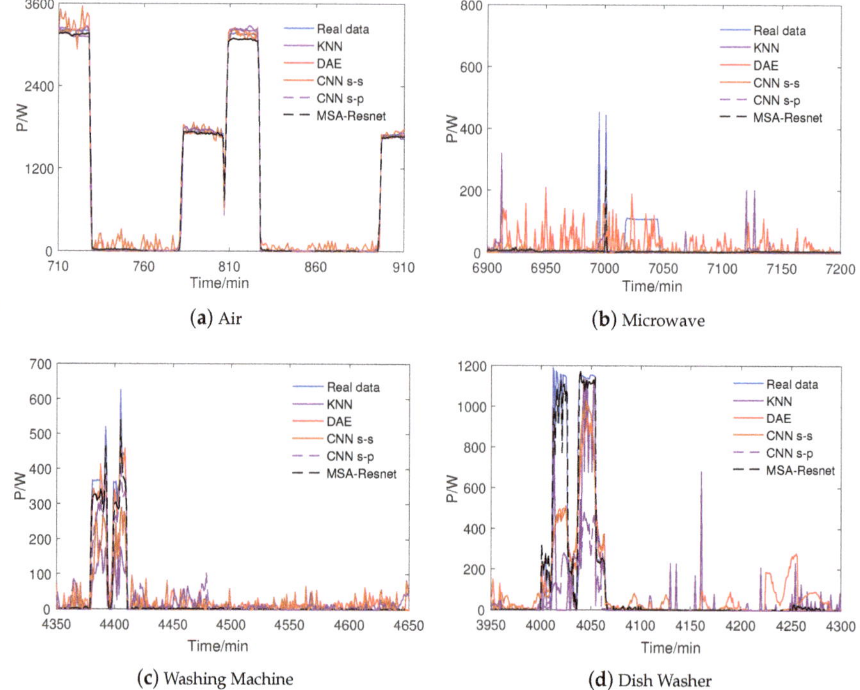

Figure 8. Load disaggregation comparison for Household 25 of the WikiEnergy dataset.

Table 1. Comparison of load disaggregation index of Household 25 of the WikiEnergy dataset.

Index	Method	Air	Fridge	Microwave	Washing Machine	Dish Washer
MAE	KNN	38.484	34.014	6.928	6.677	10.630
	DAE	36.964	39.520	17.015	12.081	25.107
	CNN s-s	61.129	38.413	9.973	18.497	19.084
	CNN s-p	39.635	13.760	13.155	11.959	11.624
	MSA-Resnet	36.388	10.440	4.862	2.161	2.013
SAE	KNN	0.0006	0.026	0.060	0.323	0.121
	DAE	0.0001	0.071	2.317	2.835	1.405
	CNN s-s	0.013	0.051	0.060	3.925	0.886
	CNN s-p	0.006	0.074	0.319	2.467	0.098
	MSA-Resnet	0.014	0.025	0.143	0.152	0.052

After the disaggregation of load, power thresholds of electrical appliances were used to distinguish the on/off states, so as to calculate their evaluation indexes. The thresholds of the air conditioner, fridge, microwave, washing machine, and dishwasher were set to 100 W, 50 W, 200 W, 20 W, and 100 W, respectively. *Recall* rate, *precision* rate, *accuracy* rate, and $F1$ values [41] were used to further evaluate the performance of the different algorithms in their on/off states.

Recall represents the probability of predicting correctly in the instance with a positive label:

$$Recall = \frac{TP}{TP + FN}, \tag{15}$$

where True Positive (TP) represents the number of predicted states that are disaggregated as "on" when their ground truth state is "on", and False Negative (FN) denotes the number of predicted states that are "on" when their ground truth state is "off". There are two possibilites: one is to predict the original positive class as a positive class (TP), and the other is to predict the original positive class as a negative class (FN).

Precision refers to the proportion of samples that are predicted to be in an "on" state and are indeed in an "on" state:

$$Precision = \frac{TP}{TP + FP}, \tag{16}$$

where False Positive (FP) represents the number of states that are actually "off" when their predicted states are "on". *Accuracy* refers to the ratio of the number of samples correctly predicted to the number of the total dataset:

$$Accuracy = \frac{TP + TN}{P + N}, \tag{17}$$

where P is the number of positive samples, and N is the number of negative samples. $F1$ can be expressed as

$$F1 = 2 \times \frac{precision \times recall}{precision + recall}. \tag{18}$$

Table 2 is a comparison of the evaluation indexes for judging "on" or "off" states of Household 25 electrical appliances. It can be seen from Table 2 that for *Accuracy* and $F1$, MSA-Resnet achieves the best performance in various electrical appliances.The disaggregation diagrams of the microwave, the washing machine, and the dishwasher are in Figure 8, which shows that, in the actual power consumption of these three electrical appliances, their proportion of "on" states is significantly lower than that of the first two electrical appliances. In such unbalanced sample data with a small sample size, the "on" states of the washing machine cannot be effectively predicted using the CNN s-s and CNN s-p, whereas the MSA-Resnet presents better results.

Table 2. Performance comparison of different algorithms for electrical on/off judgement of Household 25 in the WikiEnergy dataset.

Index	Method	Air	Fridge	Microwave	Washing Machine	Dish Washer
Recall	KNN	0.998	0.996	0.759	0.290	0.561
	DAE	0.999	0.996	1	0.451	0.833
	CNN s-s	0.999	0.990	0.949	0.290	0.868
	CNN s-p	0.999	1	0.987	0.129	0.596
	MSA-Resnet	1	0.986	0.880	0.806	1
Precision	KNN	0.987	0.870	0.198	0.236	0.336
	DAE	0.987	0.853	0.050	0.229	0.281
	CNN s-s	0.939	0.847	0.033	0.428	0.391
	CNN s-p	0.995	0.996	0.050	0.047	0.414
	MSA-Resnet	0.999	0.988	0.795	0.962	0.884
Accuracy	KNN	0.991	0.889	0.967	0.993	0.978
	DAE	0.991	0.872	0.812	0.992	0.967
	CNN s-s	0.958	0.864	0.729	0.995	0.978
	CNN s-p	0.997	0.980	0.816	0.986	0.982
	MSA-Resnet	0.999	0.981	0.997	0.999	0.998
F1	KNN	0.993	0.928	0.314	0.260	0.421
	DAE	0.993	0.919	0.095	0.304	0.421
	CNN s-s	0.968	0.913	0.064	0.346	0.539
	CNN s-p	0.997	0.986	0.096	0.069	0.489
	MSA-Resnet	0.999	0.987	0.835	0.877	0.938

In order to prove the effectiveness of the Leaky-Relu function, under the same conditions, comparative experiments are conducted with WikiEnergy's Household 25 using the Relu function. According to the experimental results in Table 3, at the two indicators of the MAE and the SAE, the algorithm using the Leaky-Relu function is better.

Table 3. Comparison of activation function of Household 25 of the WikiEnergy dataset.

Index	Function	Air	Fridge	Microwave	Washing Machine	Dish Washer
MAE	Relu	46.029	10.799	6.954	3.403	4.539
	Leaky-Relu	36.388	10.440	4.862	2.161	2.013
SAE	Relu	0.015	0.034	0.234	0.287	0.157
	Leaky-Relu	0.014	0.025	0.143	0.152	0.052

For further verification, we selected five electric appliances from Household 5 in the UK-DALE dataset for additional experiments. Figure 9 shows the results of disaggregation. The figure shows that all of the above algorithms are able to achieve effective disaggregation for the kettle, an electrical appliance that is used often. For the fridge, the KNN and the DAE work worse than the CNN s-s, the CNN s-p, and the MSA-Resnet. For the microwave, the washing machine, and the dishwasher, which are infrequently used and have a low power consumption, the MSA-Resnet has better disaggregation results than the other two deep learning algorithms, mainly because it could better detect peaks and state changes.

Table 4 shows Household 5's load disaggregation evaluation index in the UK-DALE dataset. Table 4 shows that the MSA-Resnet does better in MAE and SAE compared with other methods. For the MAE, the MSA-Resnet performs better with respect to the kettle, the fridge, the washing machine, and the dishwasher. The MSA-Resnet has smaller SAE values in the kettle, the fridge, and the washing machine.

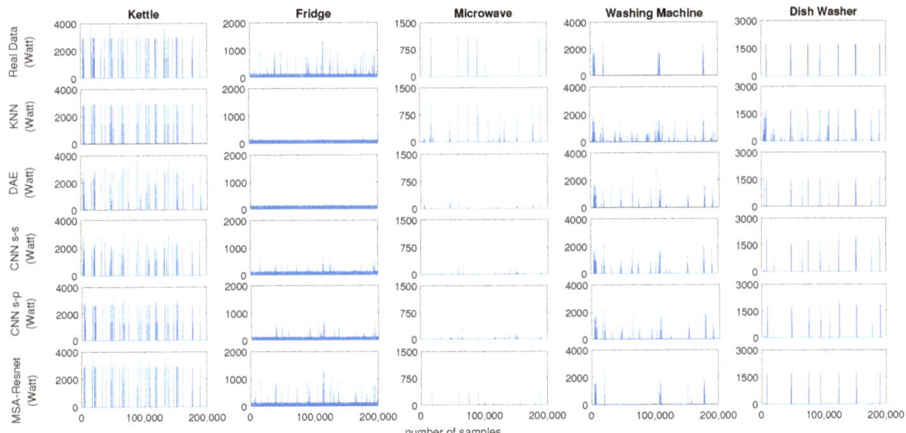

Figure 9. Comparison load disaggregation results of Household 5 in the UK-DALE dataset.

Table 4. Comparison of the load disaggregation indexes of Household 5 of the UK-DALE dataset.

Index	Method	Kettle	Fridge	Microwave	Washing Machine	Dish Washer
MAE	KNN	1.413	2.407	0.378	4.032	3.274
	DAE	8.867	8.218	1.226	14.920	12.756
	CNN s-s	8.829	3.866	1.125	20.696	9.101
	CNN s-p	4.002	4.517	1.159	23.881	9.747
	MSA-Resnet	0.804	2.136	0.906	3.618	2.601
SAE	KNN	0.076	0.015	0.054	0.018	0.001
	DAE	0.377	0.021	0.748	0.006	0.340
	CNN s-s	0.522	0.032	0.880	0.315	0.213
	CNN s-p	0.242	0.024	0.845	0.302	0.154
	MSA-Resnet	0.001	0.013	0.720	0.0007	0.050

Table 5 shows the judgement results of "on" and "off" states of Household 5 in the UK-DALE dataset. The thresholds of the kettle, the fridge, the microwave, the washing machine, and the dishwasher were set to 100 W, 50 W, 200 W, 20 W, and 100 W, respectively. Table 5 shows that the Recalls of the washing machine and the dishwasher using the CNN s-s and the CNN s-p are low, the number of positive samples is small, and its ability to predict the "on" state is poor. If the task of judging the electrical state is considered as classification, appliances with a high utilization rate have better classification results.

Figure 10 shows load disaggregation comparisons of these five methods over a period of time. It can be seen from the figure that, compared with other algorithms, the MSA-Resnet could better disaggregate equipments, whereas the KNN and the DAE have the worst decomposition abilities. For the low-frequency washing machine and dishwasher, the MSA-Resnet could still well fit the power curve, because of its network structure. It uses multi-scale convolutions to obtain rich load characteristics, and it improves the performance of the network through the attention mechanism and the residual structure.

Table 5. Performance comparison of different algorithms for electrical on/off judgement of Household 5 in the UK-DALE dataset.

Index	Method	Kettle	Fridge	Microwave	Washing Machine	Dish Washer
Recall	KNN	0.987	0.988	0.944	0.911	0.968
	DAE	0.985	0.944	0	0.921	0.938
	CNN s-s	0.969	0.990	0	0.857	0.904
	CNN s-p	0.993	0.923	0	0.838	0.928
	MSA-Resnet	1	0.994	0.951	0.927	0.946
Precision	KNN	0.998	0.974	0.933	0.617	0.799
	DAE	0.650	0.932	0	0.471	0.813
	CNN s-s	0.946	0.944	0	0.663	0.835
	CNN s-p	1	0.968	0	0.701	0.829
	MSA-Resnet	0.996	0.996	1	0.672	0.850
Accuracy	KNN	0.999	0.986	0.999	0.981	0.996
	DAE	0.997	0.955	0.999	0.967	0.996
	CNN s-s	0.999	0.975	0.999	0.983	0.996
	CNN s-p	0.999	0.961	0.999	0.984	0.996
	MSA-Resnet	1	0.996	0.999	0.985	0.997
F1	KNN	0.992	0.981	0.939	0.736	0.875
	DAE	0.783	0.938	Nan	0.623	0.871
	CNN s-s	0.957	0.967	Nan	0.748	0.868
	CNN s-p	0.996	0.945	Nan	0.764	0.876
	MSA-Resnet	0.998	0.995	0.975	0.779	0.895

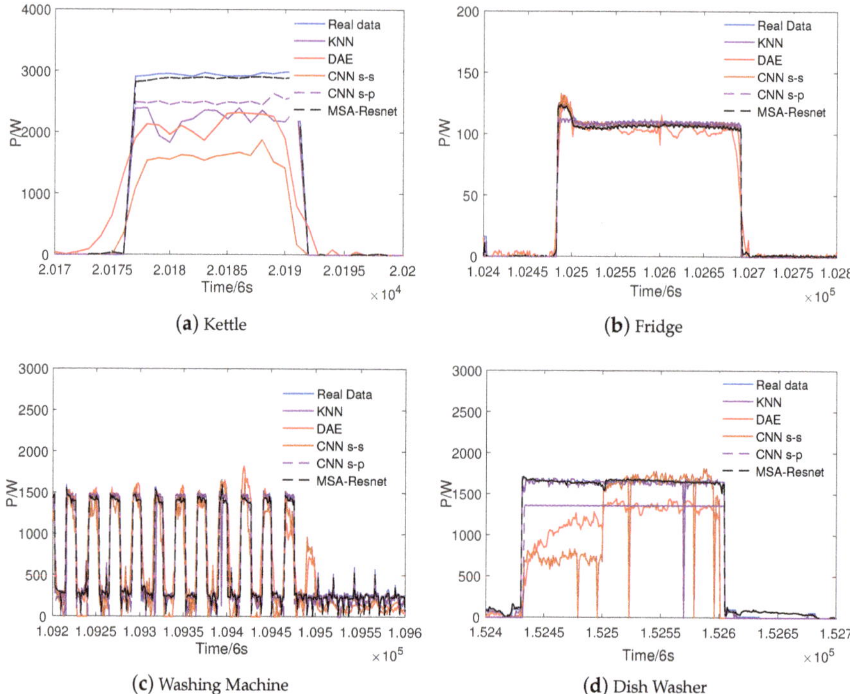

Figure 10. Load disaggregation comparison for Household 5 of the UK-DALE dataset.

In order to prove the effectiveness of the Leaky-Relu function, a comparative experiment with the Relu function was also done on the UK-DALE dataset. Table 6 can prove that the Leaky-Relu function is still the best.

Table 6. Comparison of the activation function of Household 5 of the UK-DALE dataset.

Index	Function	Kettle	Fridge	Microwave	Washing Machine	Dish Washer
MAE	Relu	2.449	4.506	1.371	25.126	4.706
	Leaky-Relu	0.804	2.136	0.906	3.618	2.601
SAE	Relu	0.286	0.061	0.950	0.253	0.036
	Leaky-Relu	0.001	0.013	0.720	0.0007	0.050

5. Conclusions

Load disaggregation is an important part of smart grids. At present, existing non-intrusive load disaggregation methods based on deep learning have some problems; for example, they easily lose features and have difficulty in detection, they do not identify low-use electrical appliances well, and their networks degrade easily due to gradient disappearance. The disaggregation effects of traditional methods are also very poor. In order to solve these problems, the MSA-Resnet is proposed for NILD. The residual network deepens the network structure, avoids the gradient, and reduces the optimization difficulty. Multi-scale convolutions obtain richer load characteristics and avoids feature simplification. The attention mechanism is used to enhance the ability of the network to learn load characteristics and improves the performance of the network. With its excellent performance on the WikiEnergy and UK-DALE datasets, the MSA-Resnet is shown to be an effective way to solve non-intrusive load disaggregation. In future work, we will conduct further experiments on public datasets such as REDD and real household data of the State Grid to verify the generalization performance of the model.

Author Contributions: Conceptualization, L.W., X.Z., and J.Q.; methodology, M.X. and Y.X.; software, K.W.; validation, X.Z. and M.X.; formal analysis, L.W., X.Z., and J.Q.; investigation, X.Z. and M.X.; resources, M.X. and K.W.; data curation, X.Z. and M.X.; writing—original draft preparation, X.Z.; writing—review and editing, L.W. and M.X.; visualization, X.Z.; supervision, M.X. and Y.X.; project administration, L.W.; funding acquisition, M.X. All authors have read and agreed to the published version of the manuscript.

Funding: This research received no external funding.

Acknowledgments: This work was supported by the State Grid Corporation of China Project 'Fundamental Theory of Dynamic Demand Response Control Based on Large-Scale Diversified Demand Side Resources'.

Conflicts of Interest: The authors declare that there is no conflict of interest regarding the publication of this paper.

References

1. Wong, Y.F.; Şekercioğlu, Y.A.; Drummond, T.; Wong, V.S. Recent approaches to non-intrusive load monitoring techniques in residential settings. In Proceedings of the 2013 IEEE Computational Intelligence Applications in Smart Grid, Singapore, 16–19 April 2013; pp. 73–79.
2. Prada, J.; Dorronsoro, J.R. General noise support vector regression with non-constant uncertainty intervals for solar radiation prediction. *J. Mod. Power Syst. Clean Energy* **2018**, *6*, 268–280. [CrossRef]
3. Liu, H.; Zeng, P.; Guo, J.; Wu, H.; Ge, S. An optimization strategy of controlled electric vehicle charging considering demand side response and regional wind and photovoltaic. *J. Mod. Power Syst. Clean Energy* **2015**, *3*, 232–239. [CrossRef]
4. Tostado-Véliz, M.; Arévalo, P.; Jurado, F. A Comprehensive Electrical-Gas-Hydrogen Microgrid Model for Energy Management Applications. *Energy Convers. Manag.* **2020**, *228*, 113726. [CrossRef]
5. Hart, G.W. Nonintrusive appliance load monitoring. *Proc. IEEE* **1992**, *80*, 1870–1891. [CrossRef]
6. Rahimpour, A.; Qi, H.; Fugate, D.; Kuruganti, T. Non-intrusive energy disaggregation using non-negative matrix factorization with sum-to-k constraint. *IEEE Trans. Power Syst.* **2017**, *32*, 4430–4441. [CrossRef]
7. Zoha, A.; Gluhak, A.; Imran, M.A.; Rajasegarar, S. Non-intrusive load monitoring approaches for disaggregated energy sensing: A survey. *Sensors* **2012**, *12*, 16838–16866. [CrossRef]

8. Chang, H.H.; Lin, L.S.; Chen, N.; Lee, W.J. Particle-swarm-optimization-based nonintrusive demand monitoring and load identification in smart meters. *IEEE Trans. Ind. Appl.* **2013**, *49*, 2229–2236. [CrossRef]
9. Lin, Y.H.; Tsai, M.S. Development of an improved time–frequency analysis-based nonintrusive load monitor for load demand identification. *IEEE Trans. Instrum. Meas.* **2013**, *63*, 1470–1483. [CrossRef]
10. Piga, D.; Cominola, A.; Giuliani, M.; Castelletti, A.; Rizzoli, A.E. Sparse optimization for automated energy end use disaggregation. *IEEE Trans. Control Syst. Technol.* **2015**, *24*, 1044–1051. [CrossRef]
11. Batra, N.; Singh, A.; Whitehouse, K. Neighbourhood nilm: A big-data approach to household energy disaggregation. *arXiv* **2015**, arXiv:1511.02900.
12. Tsai, M.S.; Lin, Y.H. Modern development of an adaptive non-intrusive appliance load monitoring system in electricity energy conservation. *Appl. Energy* **2012**, *96*, 55–73. [CrossRef]
13. Kolter, J.; Batra, S.; Ng, A. Energy disaggregation via discriminative sparse coding. *Adv. Neural Inf. Process. Syst.* **2010**, *23*, 1153–1161.
14. Johnson, M.J.; Willsky, A.S. Bayesian nonparametric hidden semi-Markov models. *J. Mach. Learn. Res.* **2013**, *14*, 673–701.
15. Kim, H.; Marwah, M.; Arlitt, M.; Lyon, G.; Han, J. Unsupervised disaggregation of low frequency power measurements. In Proceedings of the 2011 SIAM International Conference on Data Mining, Society for Industrial and Applied Mathematics, Mesa, AZ, USA, 28–30 April 2011; pp. 747–758.
16. Saitoh, T.; Osaki, T.; Konishi, R.; Sugahara, K. Current sensor based home appliance and state of appliance recognition. *SICE J. Control Meas. Syst. Integr.* **2010**, *3*, 86–93. [CrossRef]
17. Hassan, T.; Javed, F.; Arshad, N. An empirical investigation of VI trajectory based load signatures for non-intrusive load monitoring. *IEEE Trans. Smart Grid* **2013**, *5*, 870–878. [CrossRef]
18. Xia, M.; Wang, K.; Zhang, X.; Xu, Y. Non-intrusive load disaggregation based on deep dilated residual network. *Electr. Power Syst. Res.* **2019**, *170*, 277–285. [CrossRef]
19. Xia, M.; Zhang, X.; Weng, L.; Xu, Y. Multi-Stage Feature Constraints Learning for Age Estimation. *IEEE Trans. Inf. Forensics Secur.* **2020**, *15*, 2417–2428. [CrossRef]
20. Kuo, P.H.; Huang, C.J. A high precision artificial neural networks model for short-term energy load forecasting. *Energies* **2018**, *11*, 213. [CrossRef]
21. Kelly, J.; Knottenbelt, W. Neural nilm: Deep neural networks applied to energy disaggregation. In Proceedings of the 2nd ACM International Conference on Embedded Systems for Energy-Efficient Built Environments, Seoul, Korea, 4–5 November 2015; pp. 55–64.
22. Zhang, C.; Zhong, M.; Wang, Z.; Goddard, N.; Sutton, C. Sequence-to-point learning with neural networks for nonintrusive load monitoring. *arXiv* **2016**, arXiv:1612.09106.
23. Liu, Y.; Liu, Y.; Liu, J.; Li, M.; Ma, Z.; Taylor, G. High-performance predictor for critical unstable generators based on scalable parallelized neural networks. *J. Mod. Power Syst. Clean Energy* **2016**, *4*, 414–426. [CrossRef]
24. Yang, Y.; Zhong, J.; Li, W.; Gulliver, T.A.; Li, S. Semi-Supervised Multi-Label Deep Learning based Non-intrusive Load Monitoring in Smart Grids. *IEEE Trans. Ind. Inform.* **2019**, *16*, 6892–6902. [CrossRef]
25. Yadav, A.; Sinha, A.; Saidi, A.; Trinkl, C.; Zörner, W. NILM based Energy Disaggregation Algorithm for Dairy Farms. In Proceedings of the 5th International Workshop on Non-Intrusive Load Monitoring, Yokohama, Japan, 18 November 2020; pp. 16–19.
26. Faustine, A.; Pereira, L.; Bousbiat, H.; Kulkarni, S. UNet-NILM: A Deep Neural Network for Multi-tasks Appliances State Detection and Power Estimation in NILM. In Proceedings of the 5th International Workshop on Non-Intrusive Load Monitoring, Yokohama, Japan, 18 November 2020; pp. 84–88.
27. Jin, Y.; Guo, H.; Wang, J.; Song, A. A Hybrid System Based on LSTM for Short-Term Power Load Forecasting. *Energies* **2020**, *13*, 6241. [CrossRef]
28. de Paiva Penha, D.; Castro, A.R.G. Home appliance identification for NILM systems based on deep neural networks. *Int. J. Artif. Intell. Appl.* **2018**, *9*, 69–80. [CrossRef]
29. He, K.; Zhang, X.; Ren, S.; Sun, J. Deep residual learning for image recognition. In Proceedings of the IEEE Conference on Computer Vision and Pattern Recognition, Las Vegas, NV, USA, 27–30 June 2016; pp. 770–778.
30. Wang, F.; Jiang, M.; Qian, C.; Yang, S.; Li, C.; Zhang, H.; Wang, X.; Tang, X. Residual attention network for image classification. In Proceedings of the IEEE Conference on Computer Vision and Pattern Recognition, Honolulu, HI, USA, 21–26 July 2017; pp. 3156–3164.
31. Hu, J.; Shen, L.; Sun, G. Squeeze-and-excitation networks. In Proceedings of the IEEE Conference on Computer Vision and Pattern Recognition, Salt Lake City, UT, USA, 18–22 June 2018; pp. 7132–7141.

32. Vaswani, A.; Shazeer, N.; Parmar, N.; Uszkoreit, J.; Jones, L.; Gomez, A.N.; Kaiser, Ł.; Polosukhin, I. Attention is all you need. *Adv. Neural Inf. Process. Syst.* **2017**, *30*, 5998–6008.
33. Szegedy, C.; Liu, W.; Jia, Y.; Sermanet, P.; Reed, S.; Anguelov, D.; Erhan, D.; Vanhoucke, V.; Rabinovich, A. Going deeper with convolutions. In Proceedings of the IEEE Conference on Computer Vision and Pattern Recognition, Boston, MA, USA, 7–12 June 2015; pp. 1–9.
34. Lin, M.; Chen, Q.; Yan, S. Network in network. *arXiv* **2013**, arXiv:1312.4400.
35. Wang, B.; Li, T.; Huang, Y.; Luo, H.; Guo, D.; Horng, S.J. Diverse activation functions in deep learning. In Proceedings of the 2017 12th International Conference on Intelligent Systems and Knowledge Engineering (ISKE), Nanjing, China, 24–26 November 2017; pp. 1–6.
36. Nalmpantis, C.; Vrakas, D. Machine learning approaches for non-intrusive load monitoring: From qualitative to quantitative comparation. *Artif. Intell. Rev.* **2019**, *52*, 217–243. [CrossRef]
37. Kelly, J.; Batra, N.; Parson, O.; Dutta, H.; Knottenbelt, W.; Rogers, A.; Singh, A.; Srivastava, M. Nilmtk v0.2: A non-intrusive load monitoring toolkit for large scale data sets: Demo abstract. In Proceedings of the 1st ACM Conference on Embedded Systems for Energy-Efficient Buildings, Memphis, TN, USA, 4–6 November 2014; pp. 182–183.
38. Biansoongnern, S.; Plungklang, B. Non-intrusive appliances load monitoring (nilm) for energy conservation in household with low sampling rate. *Procedia Comput. Sci.* **2016**, *86*, 172–175. [CrossRef]
39. Xia, M.; Wang, K.; Song, W.; Chen, C.; Li, Y. Non-intrusive load disaggregation based on composite deep long short-term memory network. *Expert Syst. Appl.* **2020**, *160*, 113669. [CrossRef]
40. Krystalakos, O.; Nalmpantis, C.; Vrakas, D. Sliding window approach for online energy disaggregation using artificial neural networks. In Proceedings of the 10th Hellenic Conference on Artificial Intelligence, Patras, Greece, 9–12 July 2018; pp. 1–6.
41. Xia, M.; Tian, N.; Zhang, Y.; Xu, Y.; Zhang, X. Dilated multi-scale cascade forest for satellite image classification. *Int. J. Remote Sens.* **2020**, *41*, 7779–7800. [CrossRef]

Article

Evaluation of Energy Efficiency in Thermally Improved Residential Buildings, with a Weather Controlled Central Heating System. A Case Study in Poland

Krzysztof Cieśliński [1], Sylwester Tabor [2] and Tomasz Szul [2,*]

[1] Łomża Housing Cooperative in Łomża, 18-400 Łomża, Poland; kcieslinski@gmail.com
[2] Faculty of Production and Power Engineering, University of Agriculture, 30-149 Kraków, Poland; s.tabor@urk.edu.pl
* Correspondence: t.szul@urk.edu.pl; Tel.: +48-12-662-4647

Received: 5 November 2020; Accepted: 24 November 2020; Published: 26 November 2020

Abstract: Optimization of energy consumption and related energy efficiency can be realized in various ways, both through measures to reduce heat losses through building partitions and the introduction of modern systems of regulation and management of heat distribution. In order to achieve the best possible results, these actions should be interlinked, especially in older buildings that have undergone thermomodernization. Therefore, the aim of the study was to evaluate actions aimed at improving energy efficiency of buildings made in prefabricated technology. These buildings were thermomodernized and then the weather-controlled central heating system was installed. The study assessed whether the application of the change of the method of central heating regulation from the traditional one, taking into account only the change of external temperature to the weather-controlled one, will contribute to the increase of energy efficiency of buildings. The research was carried out in the existing residential buildings, for which data on the actual energy consumption was collected and elaborated and includes periods before modernization, after thermomodernization and the period after the introduction of the central heating system with weather control. The collected data cover an eighteen-year period of buildings' use. The obtained results indicate that in Polish conditions the introduction of weather-controlled regulation system in buildings made in prefabricated technology (made of large slab) allows to achieve energy savings in the range of 16–23%, it may be related to their high thermal capacity resulting from the use of concrete elements in the building envelope.

Keywords: weather controlled central system; energy saving; energy consumption; thermal improved of buildings; new energy technologies; sustainable buildings

1. Introduction

Heat is one of the main factors influencing the comfort of living or staying in the premises of residential buildings as well as public utility. It affects not only our well-being, efficiency and effectiveness but also the financial aspect related to the need to rationalize the costs of property maintenance in the phase of its exploitation and the impact on environmental protection by reducing carbon dioxide emissions to the atmosphere [1]. Single- and multi-family houses but also public buildings account on average for about 41% of total energy consumption in the European Union. These estimates reinforce the thesis that the need to reduce energy consumption in buildings, especially for heating and air conditioning, is of great importance for rational energy management. According to the report of the European Environment Agency in the European Union, energy used for space heating represents 68% of global energy consumption in buildings [2]. In Poland it is 71% [3]. For the needs of hot water, 13% is used, for lighting and other electrical appliances - 14% and only 5% for preparing meals [2].

Potential savings resulting from thermal upgrading of buildings are estimated at the level [3]:

- 33–60% to reduce energy consumption by improving thermal insulation of walls,
- 16–21% for the modernization of the ventilation system,
- 14–20% to improve the thermal insulation of the window joinery,
- 10–12% for regular inspections and modernization of central heating boilers.

The above-mentioned activities, in order to achieve the assumed effect, must take into account the so-called human factor, namely, additionally the behavior, attitudes and habits of residents- users of buildings- must change. Changing their attitude towards energy-saving measures, such as, for example: room temperature control (by means of thermostatic valves) depending on needs, limiting excessive ventilation of rooms and others, can bring real savings of 5% to 15% [4].

In recent years, a huge progress has been observed in technological solutions used in residential and commercial buildings. Starting from architectural solutions up to heating technology. More and more complicated technical innovations are being applied nowadays, whose task is to improve the living comfort in buildings. Intelligent buildings, with the help of innovative systems, have become not only more comfortable but also safer. The rapidly developing technologies make it possible to apply these modern solutions in buildings where people stay every day. Optimization of energy consumption and related energy efficiency are increasingly often analyzed in depth in practically all aspects of life of modern society. These issues, in a special way, also affect buildings and their infrastructure, because they are one of the most energy-intensive areas of human functioning. Users of residential units but also of public buildings, attach increasing importance to the possibility of offering high comfort in them and also want to make widespread use of the achievements of modern heating, ventilation or multimedia technology. The research conducted on the quality of the internal environment shows that users of various types of buildings—including thermally improved buildings—feel satisfied with their living conditions, which has a positive effect on health and productivity and at the same time translates into ecological attitudes [5–8]. This situation determines the need for a new approach to the very design phase of the planned facilities and in the case of existing buildings, their modernization and adaptation to contemporary trends and requirements [9]. These actions are taken not only in the scope of improving the comfort of use but also energy efficiency and, what is important, safety. Therefore, it has become necessary to conduct research in the scope of already available and search for new, functional building automation systems. All this in order to continuously optimize the consumption of electricity, heat and other media in buildings [10]. The most common form of energy efficiency improvement of existing buildings is their thermal improvement- often called thermomodernization. Thermomodernization is a broad concept, often mistakenly associated only with the thermal insulation of building walls. These activities also include modernization and replacement of heat sources, replacement of windows and modernization of internal heating systems. In Polish conditions, according to the research carried out on a group of 109 multi-family residential buildings, for which energy characteristics were determined, based on actual energy consumption before and after thermal improvement, it was indicated that thanks to comprehensive thermomodernization, savings of 38–40% can be achieved [11], such a level of savings can be described as "average" [12,13]. Therefore, it is advisable to use other instruments to improve energy efficiency, through modernization and improvement of heat management- operational regulation in heating systems of buildings.

The main task of operating control of each heating system is to maintain the indoor air temperature within the assumed tolerance range. A commonly used method of weather regulation of heating centers is the so called "tracking regulation," where regardless of the outside air temperature fluctuations, the regulation system maintains a constant temperature in the rooms of the heated building. There are two concepts to be distinguished here [14]:

- quantitative regulation, consisting in adjusting, depending on the individual thermal needs of the user of the premises, the amount of the flowing heating medium in the system without changing its temperature, e.g., by an appropriate setting on the thermostatic valve head,
- quality regulation, which consist in changing the temperature of the heating medium with its constant flow through the installation.

2. Literature Review and Justification of Topic Selection

In the literature can find several approaches related to the control of heating systems in buildings, which are connected to the municipal heating network.

Detailed approaches can be divided into [14,15]:

- control of night-time temperature reduction in the heating system [9],
- control based on outside temperature [15],
- energy control with indoor temperature sensors [16–18],
- control based on weather forecasts [16,19–25],
- forecast control by using a barometer [15,16],
- energy limitation by installing an accumulator tank [26,27],
- complex object control models- predictive models based on data analysis using methods such as artificial neural networks and others [21,28–32].

Some of the above-mentioned ways of regulating the operation of the heating system have been commonly used in real residential buildings as commercial systems. These are among others: EnReduce, Egain Edge oraz Kabona (these systems are discussed in detail in the paper [16]). Therefore, in this research we want to focus on the evaluation of one of the presented methods, which has been implemented in buildings in Poland.

Previous research carried out mainly in Sweden has shown that depending on the applied method of regulation, measurable savings in energy consumption can be achieved. The level of savings for selected regulation systems [16,17,20] is presented in Table 1.

Table 1. Comparison of energy saving levels for different control systems for central heating in residential buildings.

	Heating Control System in Objects		
Specification	**Egain Edge**	**EnReduce**	**Kabona**
number of buildings tested	19; 2	7	4
range of energy savings achieved %	0–26; 13,7–14	7–20	9–16
average energy saving %	11; 13,85	13	13

The presented research results indicate that the average energy savings regardless of the applied regulation system is about 13%, with the biggest differences in the achieved savings (0–26%) being observed in the case of the Egain Edge method [16]. In addition to temperature indicators (and others), one of the key values for energy storage in buildings is the time constant τ. The time constant describes how quickly a building will react to changes in the weather and additional heat. The constant differs between buildings, the heavier the structure, the greater the time constant (temperature inertia) of the building. It is estimated that a "heavy" type building has a time constant of about 160 h, an "average" building has about 66 h and a "light" building has about 25 h [33]. In the literature [16,17] there is no precise information on what type of buildings were tested, looking at the range of energy savings (0 to 26%) in the case of the method based on the Egain Edge weather forecast, it can be assumed that these buildings may have differed from each other by the type of construction and therefore by the time constant.

In Poland, from the sixties to the eighties of the last century, the technology of large plate (prefabricated elements) dominated in the residential construction. The issue of energy efficiency of large panel buildings is discussed in many scientific works [34,35]. None of the buildings erected at that time in the large panel technology, by assumption, is able to meet the thermal and energy requirements set today. The research carried out on the analyses of thermal insulation of external walls of buildings made in the large panel technology, confirmed the deterioration of their thermal insulation, mainly due to the use of concretes with increased density. The defects and errors resulting from the execution or damage of thermal insulation layers as well as the location of thermal bridges also had an impact. The occurrence of these places was caused by insufficient thickness of thermal insulation or lack of it, which additionally favored the dampness of partitions [36]. An additional factor that contributes to the appearance of moisture in building partitions is the way the apartments are used. This is due to an improper approach to ventilation of rooms, for example, covering up ventilation grilles, which leads to the occurrence of surface condensation of water vapor and, consequently, the appearance of fungus and mold. The thermomodernization carried out by insulating walls and ceilings mostly eliminated the problem of thermal bridges, however, the installation of airtight windows in old type of buildings with the simultaneous lack of sufficient ventilation may result in the occurrence of periods with increased air humidity, which indirectly affects the deterioration of the comfort of use of apartments- especially in the autumn and winter period. Hence the need to improve the regulation of the heating system, which will allow to maintain comfort of use while optimizing energy consumption in thermally improved buildings. One of the heating regulation systems in large panel buildings, which has found application in objects previously covered by thermomodernization works is Egain Edge system. This solution takes into account the prediction of heat demand, using a constantly updated weather forecast. The system consists of weather forecast receivers, which replace the existing outdoor temperature sensors and their task is to optimize heat consumption in buildings. Additionally, the Egain Edge includes climate recorders installed in selected apartments, recording humidity and temperature. According to the distributor's information, in Poland the system operates in about 400 heat distribution centers installed in multi-family residential buildings [23]. Hence the authors' interest in this topic. Preliminary research was conducted for three representative residential buildings, which were made in various prefabricated technologies. The aim of the study was to evaluate the effects of thermomodernization activities aimed at improving thermal protection of buildings made in large plate technology. The main objective was to examine whether the application of the change of the method of central heating regulation from the traditional one, taking into account only the variable external temperature to the weather control, will increase the energy efficiency (energy savings) of the buildings covered by the analysis.

3. Materials and Methods

The research was conducted for three selected prefabricated multi-family residential buildings located in the north-eastern part of Poland in the city of Łomża. The buildings have the following location: 6 Śniadeckiego St. - Building A; 12 Kołłątaja St.- Building B and 1 Niemcewicza St.- Building C. The city is located in the area of the IV Climate Zone [37] for which the calculated outdoor temperature is −22 °C, the average annual outdoor temperature is 6.9 °C and the number of degree days in a standard heating season $HDD(t_b)_0$ = 4095.4 Kd. These buildings were thermally improved in the years 2011–2015 and then the central heating control system with weather control type Egain Egde was installed in them.

The control system is connected to the existing heating system. It can only work with water heating systems, which can be controlled from one point- the central heating node. The system uses forecast control, in which the external temperature sensor controlling traditional systems is replaced by a receiver receiving local weather forecasts (Figure 1).

Figure 1. Weather forecast receiver Egain Hub, building A at 12 Śniadeckiego St. in Łomża.

The control system takes into account such parameters as: solar radiation, sunlight angle, reflection from the ground, wind speed, wind direction, surface of windows and walls of the building, orientation and thermal inertia of the building (Table 2).

Table 2. Compilation and comparison of elements included in heating regulation systems [19].

Parameters Taken into Account by the Heating Control System		Forecasting Regulation Egain Edge	Weather Control
External factors (environmental)	outside temperature	x	x
	insolation	x	-
	wind direction and speed	x	-
	precipitation	x	-
Features of the building	location of the building in relation to the directions of the world and other buildings	x	-
	ventilation system in the building	x	-
	condition of the building's facade (glazing surface)	x	-
	amount of water in the central heating system	x	-
	shape and height of the building	x	-
Internal factors	measurement of temperature and humidity in apartments	x	-

Energy savings are made by adjusting the heating system of the building to its thermal capacity on the basis of the building time constant and weather forecasts. The system uses the equivalent temperature instead of the outside temperature to determine the supply temperature setpoint. The peaks of the heating system operation are shifted and the system can work with less and more even power, because heat energy is taken from the building structure. The system continuously evaluates the quality of the forecast and in case of excessive deviations, the system switches to control according to the outside temperature instead of the forecast.

The analyzed buildings are heated from the municipal heating network. Therefore, has been information on actual heat consumption for heating in the eighteen heating seasons covering the period before and after thermomodernization. On this basis, calculations of final energy demand for heating were made and then the energy characteristics of objects in the state before and after thermal modernization were determined. To exclude seasonal fluctuations, the actual energy consumption values obtained were converted (corrected) to standard season conditions (multi-year average). The data concerning the heating season degree days (from the years 2001–2019 and the multi annual average) based on which the calculations were carried out were taken from the climate database Institute of Meteorology and Water Management-National Research Institute (IMWM-NRI) for the Łomża region. The amount of final energy consumption was calculated using the formula:

$$Q_{K,H} = \sum_{i=1}^{n} \frac{HDD(t_b)_i}{HDD(t_b)_0} \cdot Q_{K,H_i} \cdot \frac{1}{n}, \quad (1)$$

where: $Q_{K,H}$—the final energy demand for the heating season, [kWh]; $HDD(t_b)_0$—the number of degree days in a standard heating season, [°Cd]; $HDD(t_b)_i$—the number of degree days for the "i" of this year, [°Cd]; Q_{K,H_i}—final energy consumption for heating in a measurement period for the "i" of this year, [kWh]. n-number of years of the measurement period.

Due to the different periods of thermal improvement of buildings and the introduction of a weather-controlled central heating control system. Table 3 shows the values of measuring years n for energy consumption before and after modernization.

Table 3. Number of years of the measurement period—n.

Building Condition	Building		
	A	B	C
before modernization, [year]	10	12	9
after thermal improvement, [year]	2	1	2
after the introduction weather controlled central heating system, [year]	6	5	7

The index of final energy demand for heating before and after the implementation of the improvement was calculated according to the formula:

$$FE = \frac{Q_{K,H}}{A_H}, \quad (2)$$

where: FE—index of final energy demand for heating, [kWh·m^{-2}·year^{-1}]; $Q_{K,H}$—the final energy demand for the heating season, [kWh]; A_H—calculated area of temperature-controlled rooms (heated surface), [m^2].

For individual building states the FE indicator has been defined as follows: FE_0—index of final energy demand for heating converted to the conditions of the standard heating season, before modernization, [kWh·m^{-2}·year^{-1}]; FE_1—index of final energy demand for heating converted to the conditions of the standard heating season, after thermal improved, [kWh·m^{-2}·year^{-1}]; FE_2—index of final energy demand for heating converted to the conditions of the standard heating season, after the introduction weather controlled central heating system, [kWh·m^{-2}·year^{-1}].

These buildings had energy audits prepared, on the basis of which the optimum variants of thermal modernization were selected, the partitions that should be modernized were indicated and the appropriate thicknesses of layers of thermal insulation materials were selected. Some of them are measured and others calculated, as pointed out in Table 4. The table contains, among others, such information as: the year of the building's construction, the year in which the thermal improvement was performed and the year in which the modernization of the central heating system was introduced. Moreover, the table includes the following data characterizing the buildings: area and cubic capacity, area of walls through which heat losses occur, time constant of the building, power demand for heating and index of final energy demand for heating before and after the improvement consisting in insulation of external partitions.

Table 4. Characteristics of variables influencing the energy needs of analyzed buildings.

No.	Parameter	Abbreviation	Building		
			A	B	C
1	construction year of a building, [year]	C_A	1984	1992	1994
2	year in which the thermal improvement was performed, [year]	C_{ti}	2012	2014	2011
3	year in which the installing weather controlled central heating system, [year]	C_{hs}	2013	2015	2012
4	calculated from exterior measurements the heated volume of building, [m^3]	V_e	9876	10294	18376
5	calculated from interior measurements total (net internal area), [m^2]	A_{in}	2231	1646	3572
6	calculated surface of heated floors from interior measurements, [m^2]	A_f	1576	1051	2235
7	calculated from exterior measurements surface of roof projection area (net), [m^2]	A_r	602	869	1141
8	calculated from exterior measurements total walls' surface (net) area, [m^2]	A_w	1560,1	1417,3	2762,3
9	calculated surface of floor from interior measurements (floor over basement or floor on the ground), [m^2]	A_{fl}	446	549	714
10	calculated from exterior measurements total windows area, [m^2]	A_{tw}	383,5	272,5	742,2
11	number of storeys, [pc.]	N_{Os}	5	3	5
12	number of residential flats, premises [pc.]	N_{Op}	45	30	75
13	number of living persons per building [N_b]	N_{Opb}	145	99	195
14	shape coefficient of buildings (the ratio surface to volume), [m^2·m^{-3}], [m^{-1}]	S/V_e	0,37	0,36	0,38
15	calculated thermal transmittance of walls components before modernization, [W·m^{-2}·K^{-1}]	U_{w0}	0,57	0,44	0,36
16	calculated thermal transmittance of peak walls components before modernization, [W·m^{-2}·K^{-1}]	U_{pw0}	0,56	0,49	0,33
17	calculated thermal transmittance of roof projections components before modernization, [W·m^{-2}·K^{-1}]	U_{r0}	0,49	0,2	0,18
18	calculated thermal transmittance of floor components on the ground before modernization, [W·m^{-2}·K^{-1}]	U_{g0}	3,19	3,03	3,23
19	calculated thermal transmittance of floors components (floor over basement) before modernization, [W·m^{-2}·K^{-1}]	U_{f0}	1,37	0,74	0,39
20	thermal transmittance of windows (commercial data) before modernization, [W·m^{-2}·K^{-1}]	U_{win0}	2,6	2	2
21	calculated thermal transmittance of walls components after thermal improved, [W·m^{-2}·K^{-1}]	U_{w1}	0,24	0,22	0,19
22	calculated thermal transmittance of peak walls components after thermal improved, [W·m^{-2}·K^{-1}]	U_{pw1}	0,26	0,23	0,19
23	calculated thermal transmittance of roof projections components after thermal improved, [W·m^{-2}·K^{-1}]	U_{r1}	0,18	0,2	0,18
24	calculated thermal transmittance of floor components on the ground after thermal improved, [W·m^{-2}·K^{-1}]	U_{g1}	3,19	3,03	3,23
25	calculated thermal transmittance of floors components (floor over basement) after thermal improved, [W·m^{-2}·K^{-1}]	U_{f1}	1,37	0,74	0,39
26	thermal transmittance of windows (commercial data) after thermal improved, [W·m^{-2}·K^{-1}]	U_{win1}	1,4	1,4	1,4
27	calculated time constant, [h]	τ	78	62	84
28	calculated heating consumed power, [kW]	Φ_h	154	125	236
29	calculated temperature of the internal wall surface before modernization, [°C]	T_{iwb}	16	16,91	17,47
30	calculated temperature of the internal wall surface after thermal improved, [°C]	T_{iwa}	18,32	18,46	18,67
31	measured, (average) index of final energy demand for heating converted to the conditions of the standard heating season, before modernization, [kWh·m^{-2}·year^{-1}]	$FE_{0(avg)}$	95,4	104,0	102,9
32	measured, index of final energy demand for heating converted to the conditions of the standard heating season, after thermal improved, [kWh·m^{-2}·year^{-1}]	FE_1	74	77,6	95,3

The buildings subjected to the analysis were characterized by a similar value of the unit energy demand indicator for heating, which ranged from about 95 to 104 [kWh·m^{-2}·year^{-1}]. The demand for heating power after thermal improvement is in the range from 125 kW (building B) to 236 kW (building C).

The calculated building time constants τ (according to EN 13790 [38]) range from 62 h for building B to 78 and 84 h for buildings A and C respectively. They can be classified as "medium" buildings [34], with buildings A and C having a higher thermal inertia compared to building B.

For the analyzed buildings, the amount of energy savings was calculated for two states:

(a) after thermal improvement, where the energy saving in percent is calculated according to equation:

$$FE_{\%,ES,a} = \left(1 - \frac{FE_1}{FE_0}\right)\cdot 100, \tag{3}$$

where: $FE_{\%,ES,a}$—the energy savings achieved through thermal improvement of buildings in percent; FE_0—(average) index of final energy demand for heating converted to the conditions of the standard heating season, before modernization, [kWh·m^{-2}·year^{-1}]; FE_1—index of final energy demand for heating converted to the conditions of the standard heating season, after thermal improved, [kWh·m^{-2}·year^{-1}].

(b) after the introduction weather controlled central heating system.

The energy saving in percent is calculated as:

$$FE_{\%,ES,b,i} = \left(1 - \frac{FE_{2,i}}{FE_{1,r}}\right)\cdot 100, \tag{4}$$

where: $FE_{\%,ES,b,i}$—the annual energy savings achieved through the introduction weather controlled central heating system in percent; $FE_{2,i}$—index of final energy demand for heating for the "i" of this year, converted to the conditions of the standard heating season, after the introduction weather controlled central heating system, [kWh·m^{-2}·year^{-1}]; $FE_{1,r}$—index of final energy demand for heating converted to the conditions of the standard heating season, in the reference year, [kWh·m^{-2}·year^{-1}]. The year preceding the installation of the Egain Edge system was adopted as the reference year, converted to the conditions of the standard heating season.

- Building A. As the system was installed in September 2013, the first period with a full 12-month cycle, from January to December, included in the savings analysis, is 2014.
- Building B. As the system was installed at the beginning of January 2015, the first period with a full 12-month cycle, from January to December, included in the savings analysis, is 2015.
- Building C. As the system was installed in September 2012, the first period with a full 12-month cycle, from January to December, included in the savings analysis, is 2013.

Reducing energy consumption is calculated as:

$$\Delta FE_{,ES,b,i} = FE_{1,r} - FE_{2,i}, \tag{5}$$

where: $\Delta FE_{,ES,b,i}$—the annual reducing energy consumption, [kWh·m^{-2}·year^{-1}]; $FE_{1,r}$—index of final energy demand for heating converted to the conditions of the standard heating season, in the reference year, [kWh·m^{-2}·year^{-1}]; $FE_{2,i}$—index of final energy demand for heating for the "i" of this year, converted to the conditions of the standard heating season, after the introduction weather controlled central heating system, [kWh·m^{-2}·year^{-1}].

Total energy savings as a result of comprehensive thermomodernization (wall insulation plus introduction weather controlled central heating system) for the buildings analyzed is calculated as:

$$FE_{\%,ES,tot} = \left(1 - \frac{FE_{2(avg)}}{FE_{0(avg)}}\right)\cdot 100, \tag{6}$$

where: $FE_{\%,ES,tot}$—the energy savings achieved through thermal improvement of buildings in percent; $FE_{0(avg)}$—(average) index of final energy demand for heating converted to the conditions of the standard heating season, before modernization, [kWh·m^{-2}·year^{-1}]; $FE_{2,i}$—(average) index of final energy demand for heating, converted to the conditions of the standard heating season, after the introduction weather controlled central heating system, [kWh·m^{-2}·year^{-1}].

4. Results and Discussion

The consumption of thermal energy for the analyzed buildings for the needs of central heating in the years 2001–2019 is shown in Figures 2–4. The amount of consumption was expressed using unit indicators of final energy demand, covers the periods before modernization, after thermal improvement (reference year) and the period after the introduction weather controlled central heating system (Egain).

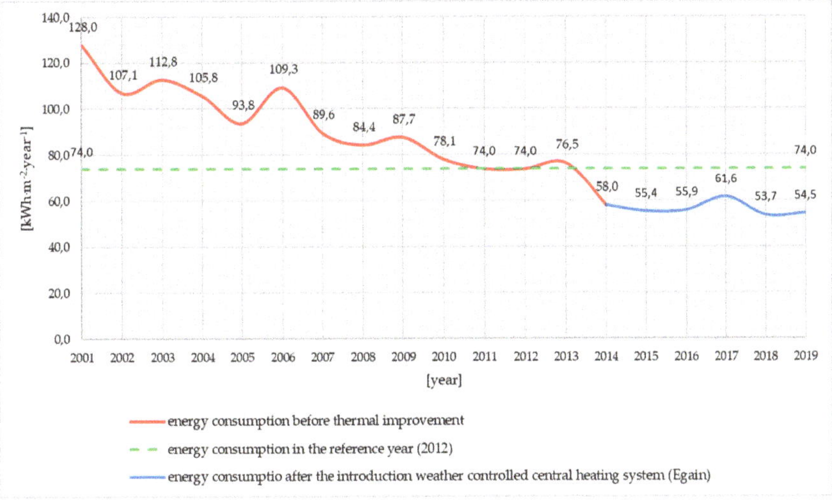

Figure 2. Thermal energy consumption for central heating in the years 2001–2019 in building A.

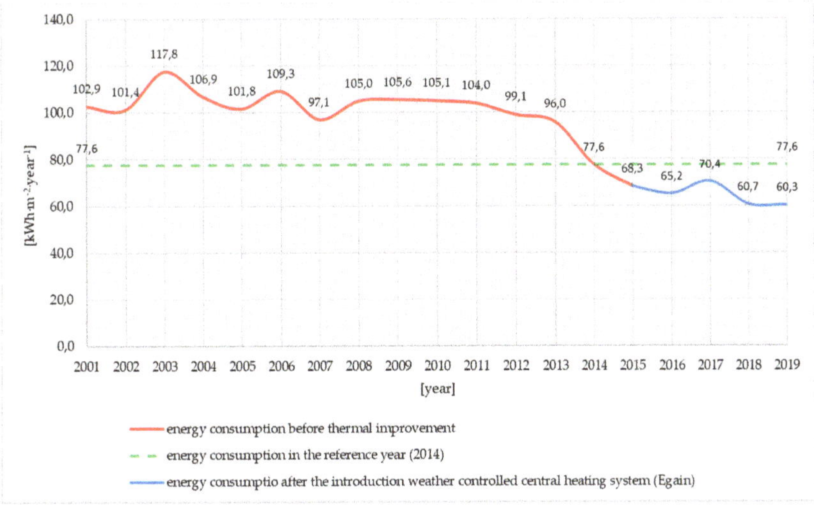

Figure 3. Thermal energy consumption for central heating in the years 2001–2019 in building B.

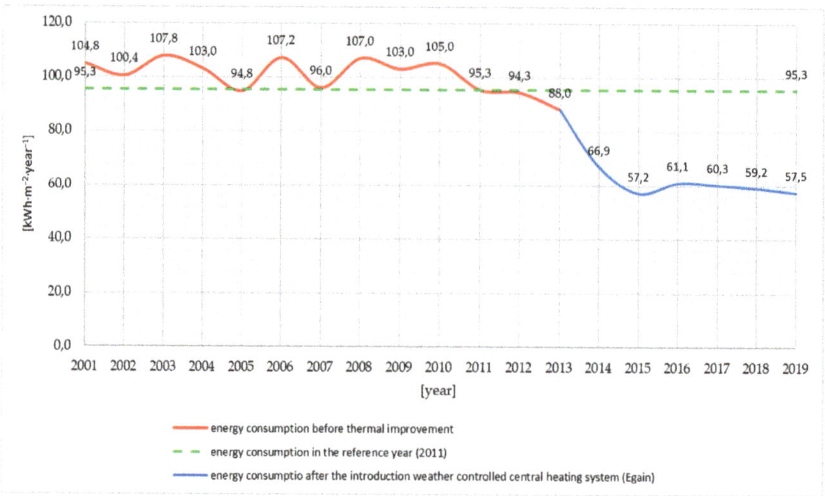

Figure 4. Thermal energy consumption for central heating in the years 2001–2019 in building C.

In the years 2011–2014, the buildings were thermomodernized by insulating the walls. After the thermal improvement (without modernization of the heating system), energy consumption was reduced, which is shown in Table 5.

Table 5. Energy savings achieved through thermal improvement of buildings.

$FE_{\%,ES,a}$	Building		
	A	B	C
savings achieved, %	22,5	25,4	7,4

The thermal improvement of the external partitions has had the best effect in buildings A and B, where energy consumption has been reduced by about 22–25%. In building C, thermal modernization of the walls did not bring much savings, because only about 7%.

The next stage of the research was to check whether the application of the change in the method of central heating regulation from the traditional one, taking into account only the change in the outside temperature to the weather control, will increase the energy efficiency (savings in energy consumption) of the buildings covered by the analysis. For this purpose, the amount of energy consumption reduction $\Delta FE_{,ES,b,i}$ and energy savings $FE_{\%,ES,b,i}$ were calculated. The calculations were made, for each (full) year after installation of the Egain control system according to formulas 4 and 5. Then the average values of the energy savings obtained were calculated. The results of the calculations are summarized in Table 6.

Table 6. The impact of the introduction weather controlled central heating system on the reduction of energy consumption in the building.

Year	Building					
	A		B		C	
	$\Delta FE_{,ES,b,i}$ [kWh·m^{-2}·Year^{-1}]	$FE_{\%,ES,b,i}$ %	$\Delta FE_{,ES,b,i}$ [kWh·m^{-2}·Year^{-1}]	$FE_{\%,ES,b,i}$ %	$\Delta FE_{,ES,b,i}$ [kWh·m^{-2}·Year^{-1}]	$FE_{\%,ES,b,i}$ %
2013					7,3	7,7
2014	16	21,6			28,4	29,9
2015	18,6	25,3	9,3	12	38,1	39,9
2016	18,1	24,5	12,4	15,8	34,2	35,9
2017	12,4	16,7	7,2	9,3	35	36,8
2018	20,3	27,6	16,9	21,8	36,1	37,9
2019	19,5	26,3	17,3	22,3	37,8	39,8
average	17,5	23,6	12,6	16,2	30,9	32,5

Analyzing the results contained in Table 6, it can be concluded that the change of the method of central heating control from the traditional one, taking into account only the external temperature variation to the weather control, has resulted in increased energy efficiency in the analyzed buildings. The energy consumption in comparison with the reference year (after the partitions have been insulated) has decreased on average from 12.6 to 30.9 kWh·m^{-2}·year^{-1}. The energy savings are about 16% to 32%. Comparing the savings associated with the introduction of a new control system for central heating installations for individual buildings in relation to the savings obtained as a result of insulating the walls, it can be concluded that in the case of building C, there is a supposition that only a 7% reduction in energy consumption after insulating the walls (where from the theoretical calculations should be about 20% [9]), could be caused by faulty operation of the central heating control system. After changing the control system to a forecast, there was a significant reduction in energy consumption, which in 2015 and 2019 reached almost 40%.

Taking into account the above observations when estimating energy efficiency, it was decided to refer to buildings A and B, where as a result of thermomodernization the obtained savings levels are similar and amount to 22% to 25%. These buildings, after thermal improvement, were characterized by similar energy consumption rates in the reference year, which amounted to 74–77,6 kWh·m^{-2}·year^{-1}. In these buildings, the change of regulation method resulted in average reduction of energy consumption in the range of 12,6 kWh·m^{-2}·year^{-1} (building B) to 17,5 kWh·m^{-2}·year^{-1} (building A). Energy saving in the range of 16,2% (building B) to 23,6% (building A). It can be assumed that the differences in achieved savings may result from different curtain wall structures, where in building A prefabricated elements based on reinforced concrete were used, while in building B prefabricated elements based on aerated concrete. The values of calculated time constants of buildings 78 h for building A and 62 h for building B also prove it.

Comparing the obtained results with the data presented in the literature, where the obtained savings amount to about 13–14% [16,17,20], it can be stated that the application of weather-controlled central heating system in buildings made in the prefabricated technology (which have been thermally improved) gives better results, which, depending on the applied technology of making curtain walls, can give the average energy savings of about 20%.

The total energy savings achieved as a result of the thermal improvement combined with the modernization of the heating system are presented in Table 7.

Table 7. Total energy savings achieved through thermal improvement of buildings and of the introduction weather controlled central heating system.

$FE_{\%,ES,tot}$	Building		
	A	B	C
total savings achieved, %	40,7	37,5	37,5

Buildings, that have undergone comprehensive thermal upgrading measures can achieve energy savings on heating of about 37% to 41%.

5. Conclusions and Perspectives

The paper presents an evaluation of measures to improve the energy efficiency of buildings made in large plate technology. These buildings were subjected to thermal improvement consisting in insulation of external partitions and then the weather controlled central heating system was installed. The main objective was to examine whether the application of the change of the method of central heating regulation from traditional, taking into account only the variation of outside temperature to weather control, will increase the energy efficiency (energy savings) of buildings. The analysis covered the period between 2001 and 2019. The amount of energy used by buildings for heating was determined on the basis of actual consumption and then converted to the conditions of the standard heating season. Annual energy consumption is presented using unit final energy demand indicators and covers the periods before modernization, after thermal modernization (reference year) and the period after introduction weather controlled central heating system. The thermal improvement of the external partitions has energy consumption has been reduced by about 22–25%. Modernization of heating consisting in the implementation of weather-controlled central heating system, depending on the technology used to make curtain walls in buildings, allows to achieve savings in the range of about 16–23%. The total energy savings in the analyzed buildings is on average about 39%.

The presented results of the research carried out for three representative residential buildings can be treated as a preliminary one, where they wanted to indicate the potential for savings that can be achieved in buildings made in the large panel technology. Further research should be carried out on a set of dozens of buildings, so that it is possible to identify specific groups of buildings, made in prefabricated technology using various materials, for which, after applying the regulation of weather controlled central heating system, it will be possible to indicate a specific level of energy savings—characteristic for a given group.

Author Contributions: Conceptualization, T.S.; data curation, K.C.; investigation, T.S. and K.C.; methodology, T.S.; project administration, T.S.; supervision, T.S.; writing—original draft, T.S.; writing—reviewing and editing, T.S; K.C. and S.T. All authors have read and agreed to the published version of the manuscript.

Funding: This research was financed by the Ministry of Science and Higher Education of the Republic of Poland.

Acknowledgments: We are grateful to Egain Polska sp. z o.o. for providing data on the energy consumption of the analyzed buildings and Urszula Malaga-Toboła from the University of Agriculture in Krakow, Poland, Faculty of Production and Power Engineering for her valuable support during this research.

Conflicts of Interest: The authors declare no conflict of interest.

References

1. Adamczyk-Królak, I. Analysis of Heat Energy Savings in Single-Family Housing. *Constr. Optim. Energy Potential (CoOEP)* **2014**, *1*, 9–14. Available online: http://www.bozpe.bud.pcz.pl/ANALYSIS-OF-HEAT-ENERGY-SAVINGS-IN-SINGLE-FAMILY-HOUSING,91052,0,2.html (accessed on 25 October 2020).
2. European Commission. Energy Consumption by End-Use in Residential Buildings. Available online: https://ec.europa.eu/energy/content/energy-consumption-end-use_en (accessed on 25 October 2020).
3. Lis, P.; Sekret, R. Efektywność Energetyczna Budynków-Wybrane Zagadnienia Problemowe. *Rynek Energii* **2016**, *6*, 29–35. Available online: http://yadda.icm.edu.pl/baztech/element/bwmeta1.element.baztech-dcd9364e-6cc9-427e-b150-3fbae739f649?q=bwmeta1.element.baztech-a4741551-e340-47a9-887d-714a9ba80911;4&qt=CHILDREN-STATELESS (accessed on 26 October 2020).
4. European Environment Agency. Achieving Energy Efficiency Through Behaviour Change: What does It Take? EEA Technical Report No 5/2013. Available online: https://www.eea.europa.eu/publications/achieving-energy-efficiency-through-behaviour/file (accessed on 25 October 2020).
5. Caniato, M.; Gasparella, A. Discriminating People's Attitude towards Building Physical Features in Sustainable and Conventional Buildings. *Energies* **2019**, *12*, 1429. [CrossRef]

6. Sant'Anna, D.O.; Dos Santos, P.; Vianna, N.; Romero, M. Indoor environmental quality perception and users' satisfaction of conventional and green buildings in Brazil. *Sustain. Cities Soc.* **2018**, *43*, 95–110. [CrossRef]
7. De La Cruz-Lovera, C.; Perea-Moreno, A.-J.; De La Cruz-Fernández, J.-L.; Montoya, F.G.; Alcayde, A.; Manzano-Agugliaro, F. Analysis of Research Topics and Scientific Collaborations in Energy Saving Using Bibliometric Techniques and Community Detection. *Energies* **2019**, *12*, 2030. [CrossRef]
8. Leaman, A.; Bordass, B. Are users more tolerant of 'green' buildings? *Build. Res. Inf.* **2007**, *35*, 662–673. [CrossRef]
9. Szul, T. *Energy Efficiency Assessment of Buildings*; Scientific Publishing House INTELLECT: Waleńczów, Poland, 2018; ISBN 978-83-950526-3-7. (In Polish)
10. Noga, M.; Ożadowicz, A.; Grela, J. Efektywność Energetyczna i Smart Metering–nowe Wyzwania dla Systemów Automatyki Budynkowej. *Napędy Sterow.* **2012**, *12*, 54–59. Available online: http://beta.nis.com.pl/userfiles/editor/nauka/122012_n/Oadowicz_12-2012.pdf (accessed on 25 October 2020).
11. Szul, T.; Kokoszka, S. Application of Rough Set Theory (RST) to Forecast Energy Consumption in Buildings Undergoing Thermal Modernization. *Energies* **2020**, *13*, 1309. [CrossRef]
12. BPIE; Staniaszek, D.; Firląg, S. Financing Building Energy Performance Improvement in Poland. StatusReport. Available online: http://bpie.eu/wp-content/uploads/2016/01/BPIE_Financing-building-energy-in-Poland_EN.pdf (accessed on 18 November 2020).
13. Adamczyk, J.; Dylewski, R. Ecological and Economic Benefits of the "Medium" Level of the Building Thermo-Modernization: A Case Study in Poland. *Energies* **2020**, *13*, 4509. [CrossRef]
14. Kathirgamanathan, A.; De Rosa, M.; Mangina, E.; Finn, D.P. Data-driven predictive control for unlocking building energy flexibility: A review. *Renew. Sustain. Energy Rev.* **2021**, *135*, 110120. [CrossRef]
15. Kasper, T. *The Energy Efficiency Potential of Intelligent Heating Control Approaches in the Residential Sector*; Eidgenössische Technische Hochschule Zürich: Zurich, Switzerland, 2013; Available online: https://sustec.ethz.ch/content/dam/ethz/special-interest/mtec/sustainability-and-technology/PDFs/130420%20Master%20thesis%20on%20intelligent%20heating%20control%20approaches%20-%20Thomas%20Kasper_final.pdf (accessed on 14 October 2020).
16. Persson, J.; Vogel, D. *Utnyttjande av Byggnaders Värmetröghet. Utvärdering av Kommersiella Systemlösningar*; Lunds Universitet: Lund, Sweden, 2011; ISRN LUTVDG/TVIT–11/5030–SE(68); Available online: http://lup.lub.lu.se/luur/download?func=downloadFile&recordOId=3161459&fileOId=3161460 (accessed on 20 October 2020).
17. Herrlin, E. *Alternativa Reglermetoder för en Energieffektiv Byggnad*; KTH Skolan för Kemi Och Hälsa: Stockholm, Sweden, 2017; Available online: http://kth.diva-portal.org/smash/get/diva2:1143765/FULLTEXT02.pdf. (accessed on 5 October 2020).
18. Enreduce Energy Control AB. Available online: https://www.enreduce.se/om-oss/ (accessed on 21 October 2020).
19. Oldewurtel, F.; Parisio, A.; Jones, C.; Gyalistras, D.; Gwerder, M.; Stauch, V.; Lehmann, B.; Morari, M. Use of model predictive control and weather forecasts for energy efficient building climate control. *Energy Build.* **2012**, *45*, 15–27. [CrossRef]
20. Hilding, O.; Nilsson, S. *Analysis and Development of Control Strategies for a District Heating Central*; Chalmers University of Technology: Gothenburg, Sweden, 2009; Available online: http://publications.lib.chalmers.se/records/fulltext/99408.pdf (accessed on 18 October 2020).
21. Cox, R.A.; Drews, M.; Rode, C.; Nielscen, S.B. Simple future weather files for estimating heating and cooling demand. *Build. Environ.* **2015**, *83*, 104–115. [CrossRef]
22. Kabona, Ecopilot. Available online: https://nordomatic.com/building-automation/ecopilot/ (accessed on 21 October 2020).
23. Egain Edge. Available online: https://www.egain.io/pl/nasz-platforma/technologia/#sztuczna-inteligencja (accessed on 21 October 2020).
24. Adamski, M.; Ruszczyk, J. New Weather Controlled Central Heating System. *Ciepłownictwo Ogrzew. Went.* **2012**, *43*, 278–283. Available online: https://www.researchgate.net/publication/299447251 (accessed on 21 October 2020).
25. Bacher, P.; Madsen, H.; Nielsen, H.A.; Perers, B. Short-term heat load forecasting for single family houses. *Energy Build.* **2013**, *65*, 101–112. [CrossRef]
26. Warfvinge, C.; Dahlblom, M. *Projektering av VVS-Installationer*; Studentlitteratur AB: Lund, Sweden, 2010; ISBN 978-914-405-5619.

27. Zemann, C.; Deutsch, M.; Zlabinger, S.; Hofmeister, G.; Gölles, M.; Horn, M. Optimal operation of residential heating systems with logwood boiler, buffer storage and solar thermal collector. *Biomass Bioenergy* **2020**, *140*, 105622. [CrossRef]
28. Huang, H.; Chen, L.; Hu, E. A neural network-based multi-zone modelling approach for predictive control system design in commercial buildings. *Energy Build.* **2015**, *97*, 86–97. [CrossRef]
29. Cigler, J.; Prívara, S.; Váňa, Z.; Žáčeková, E.; Ferkl, L. Optimization of Predicted Mean Vote index within Model Predictive Control framework: Computationally tractable solution. *Energy Build.* **2012**, *52*, 39–49. [CrossRef]
30. Fang, J.; Ma, R.; Deng, Y. Identification of the optimal control strategies for the energy-efficient ventilation under the model predictive control. *Sustain. Cities Soc.* **2020**, *53*, 101908. [CrossRef]
31. Moon, J.W. Performance of ANN-based predictive and adaptive thermal-control methods for disturbances in and around residential buildings. *Build. Environ.* **2012**, *48*, 15–26. [CrossRef]
32. Hietaharju, P.; Ruusunen, M.; Leiviskä, K. Enabling Demand Side Management: Heat Demand Forecasting at City Level. *Materials* **2019**, *12*, 202. [CrossRef]
33. Bröms, G.; Isfält, E. Effekt-och Energibesparing Genom Förenklad Styrning och Drift av Installationsystem i Byggnader. Sztokholm: Institutionen Förinstallationsteknik, KTH, ISSN: 0284-141X. 1992. Available online: https://www.kabona.com/wp-content/uploads/2015/12/Sammanfattning_teoretisk_bakgrund_Ecopilot.pdf (accessed on 17 October 2020).
34. Szul, T.; Nęcka, K.; Mathia, T. Neural Methods Comparison for Prediction of Heating Energy Based on Few Hundreds Enhanced Buildings in Four Season's Climate. *Energies* **2020**, *13*, 5453. [CrossRef]
35. Dębowski, J. Cała prawda o budynkach wielkopłytowych. *Przegląd Bud.* **2012**, *83*, 28–35.
36. Dohojda, M.; Wiśniewski, K. Termomodernizacja sposobem rewitalizacji osiedli mieszkaniowych z wielkiej płyty. *Przegląd Bud.* **2019**, *90*, 47–50.
37. CEN. *European Standard: Heating Systems in Buildings*. PN-EN ISO 12831-1:2017-08. 2017. Available online: https://sklep.pkn.pl/pn-en-12831-3-2017-08e.html (accessed on 5 October 2020).
38. CEN. *European Standard: Energy Performance of Buildings-Calculation of Energy Use for Space Heating and Cooling*. ISO 13790:2008. 2008. Available online: https://www.iso.org/standard/41974.html (accessed on 5 October 2020).

 applied sciences

Review

Hybrid Techniques to Predict Solar Radiation Using Support Vector Machine and Search Optimization Algorithms: A Review

José Manuel Álvarez-Alvarado [1], José Gabriel Ríos-Moreno [2,*], Saul Antonio Obregón-Biosca [2], Guillermo Ronquillo-Lomelí [3], Eusebio Ventura-Ramos, Jr. [2] and Mario Trejo-Perea [2]

1. División de Investigación y Posgrado, Facultad de Ingeniería, Universidad Autónoma de Queretaro, Queretaro 76010, Mexico; jmalvarez@uaq.mx
2. Facultad de Ingeniería, Universidad Autónoma de Queretaro, Queretaro 76010, Mexico; saul.obregon@uaq.mx (S.A.O.-B.); eventura@uaq.mx (E.V.-R.J.); mtp@uaq.mx (M.T.-P.)
3. Centro de Ingeniería y Desarrollo Industrial, Querétaro 76125, Mexico; gronquillo@cidesi.edu.mx

* Correspondence: riosg@uaq.mx; Tel.: +52-442-192-1200 (ext. 6064)

Abstract: The use of intelligent algorithms for global solar prediction is an ideal tool for research focused on the use of solar energy. Forecasting solar radiation supports different applications focused on the generation and transport of energy in places where there are no meteorological stations. Different solar radiation prediction techniques have been applied in different time horizons, such as neural networks (ANN) or machine learning (ML), with the latter being the most important. The support vector machine (SVM) is a classification method of the ML that is used to predict solar radiation. To obtain a better accuracy of prediction data, search optimization algorithms (SOA) such as genetic algorithms (GA) and the particle swarm optimization algorithm (PSO) were used to optimize the prediction accuracy by searching the model parameters. This article presents a review of different hybrid SVM models with SOA applied to obtain the best parameters to reduce the prediction error of solar radiation using meteorological variables. Research articles from the last 5 years on solar radiation prediction using SVM models and hybrid SMV optimized models with SOA were studied. The results show that SVM with GA presents a better performance than the classical SVM models using the Radial basis kernel function for prediction parameters.

Keywords: solar radiation; support vector machine; heuristic algorithm; renewable energy; solar energy systems

Citation: Álvarez-Alvarado, J.M.; Ríos-Moreno, J.G.; Obregón-Biosca, S.A.; Ronquillo-Lomelí, G.; Ventura-Ramos, E., Jr.; Trejo-Perea, M. Hybrid Techniques to Predict Solar Radiation Using Support Vector Machine and Search Optimization Algorithms: A Review. *Appl. Sci.* **2021**, *11*, 1044. https://doi.org/10.3390/app11031044

Academic Editor: José A. Orosa
Received: 29 November 2020
Accepted: 20 January 2021
Published: 24 January 2021

1. Introduction

In recent years, energy generation and transport have become very important issues for the social and economic development of any nation that wants to be sustainable. Today, the demand for fossil fuel is 80% of the total energy consumed globally and more than 95% is used for the transport sector [1]. The use of these fossil fuels has been one of the main causes of the greenhouse effect on earth [2–4]. Nowadays, the scientific community has set the task of developing new technologies focused on the generation and use of electricity through solar power [5–7].

Additionally, power companies must be able to manage energy production to meet consumption at any time [8]. This is why it has focused on generating new techniques to manage energy production, as it is an important factor for a society to thrive economically and without harming the environment by using alternative energies [9]. However, alternative energies (such as solar, wind, to name a few), are difficult to represent in a mathematical model because of their non-linear behavior. In order to meet the balance between generation and consumption, it is crucial to predict solar radiation in high-capacity power generation facilities.

In this context, acquiring further knowledge on solar radiation has been one of the main research topics, for which it has become a benchmark in the strengthening of energy generation strategies through the use of renewable energy sources. In this context, machine learning (ML) has become a recognized strategy in this field [10]. A high-performance solar energy generation system largely depends on the forecast of the output power, since this data can support the design and sizing of these systems. Under this concept, forecasting models of global solar radiation are developed under two main categories: satellite cloud images and ML models [11].

Forecasting solar radiation is becoming a popular topic. This technology allows solar energy to be integrated into the grid, producing good results by improving the quality of energy supplied to the grid to reduce the costs of accessories related to the use of this resource [12]. The combination of these factors has been the motivator for the development and design of models of a complex field of research that aims to produce better predictions of the solar resource and thus be able to predict the output power that can be generated depending on the type of technology used and the prediction horizon used [13].

A forecasting model essentially consists of a system of linear or non-linear equations that relate the future values of the variable to be predicted with recorded data variables themselves and the explanatory variables. Before making a prediction, you must define the prediction horizon on which the model should be applied [14]. According to the literature, there are two classes of techniques to choose the method according to horizon time: The Now-casting method, defined as a forecast for the next 6-hour period, based on detailed observational data such as radiometers, pyrheliometers, satellite images or sky cameras, among others and results in a better alternative to forecast variables in a minute scale [15] and Numeric weather prediction models (NWP) [8]. These predictions are suitable for the operation and control of power plants. For some applications, solar radiation predictions of 0 to 180 h are delivered online, every 6 h [11].

In addition, another advantage of knowing the future solar radiation lies in optimizing the control of solar energy in the electricity grid, which can ensure a favorable performance in the electricity generation market that may be used in the future in the Smart Grid field [16].

Currently, there is a lot of information published in journals about solar radiation prediction with ML. However, several ML methods about classification models to forecast variables are considered in the literature, such as artificial neural networks (ANN) or support vector machine (SVM) models and are commonly compared with other models. Additionally, forecasting solar radiation is a very complex phenomenon which can be influenced by many different factors and forecasting models tend to be more accurate. In this context, optimization algorithms are integrated to forecasting models to reduce error and improve its accuracy, so, it is necessary to present a review of prediction techniques with the use of search optimization algorithms (SOA) to improve prediction on different horizons. The objective of this article is to perform an analysis of solar radiation prediction techniques based on hybrid SVM and SOA models. The performance of SVM models is compared with conventional supervised learning models, such as artificial neural networks. SVM-SOA hybrid models are also analyzed to evaluate the accuracy of the predicted solar radiation data seen in the literature. It also presents the most suitable Kernel functions for hybrid models and finally, a general process flowchart of hybrid prediction techniques, according to the literature, is shown.

2. Solar Radiation Components

Modelling solar radiation is a very complex task because it is influenced by climatic zone, geographical area or seasons. Solar radiation provides the quantity of solar energy that reaches the Earth's surface during a particular time period [17]. It is important to know the whole phenomenon, starting from the definition of the sun, which is a star inside which a series of reactions take place that produce a loss of mass that is transformed into energy. Solar irradiance is the measured and recorded amount of energy that comes directly from the sun to the land surface [18]. There are three main types of solar radiation, which are the following: diffuse, direct and global solar radiation. Diffuse radiation is that which occurs when radiation that comes directly from the sun is intercepted by the Earth's atmosphere, causing a scattering in the middle [19]. Diffuse solar radiation creates a problem in the generation of electricity by photovoltaic solar panels, reducing their generation capacity, that is, in solar radiation, clouds absorb all the incident energy and emit it again [20]. Direct solar radiation comes in a defined direction from the sun towards the earth, which, it is possible to concentrate it in a point for its use, however, it can also be reflected. This radiation is essential in the use and sizing of solar concentration systems [21]. With diffuse and direct solar radiation, it is possible to determine the global solar radiation in an area. Furthermore, it is possible to acquire information on global solar radiation by means of pyranometers that measure the solar irradiance from the sun to an area [22].

3. Solar Radiation Prediction Time Horizons

A prediction model essentially consists of a system of linear or non-linear equations that relate the future values of the variable to be predicted with the present and past values of the variable itself and the explanatory variables. Before making a prediction, you must define the prediction horizon on which the model should be applied [14]. The ML applications to predict solar radiations are becoming a trend in the transition of energy generation systems. The models are developed as a time series prediction problem, that will be solved as a classification model. Figure 1 visualizes the most used time horizons in prediction models using ML [23].

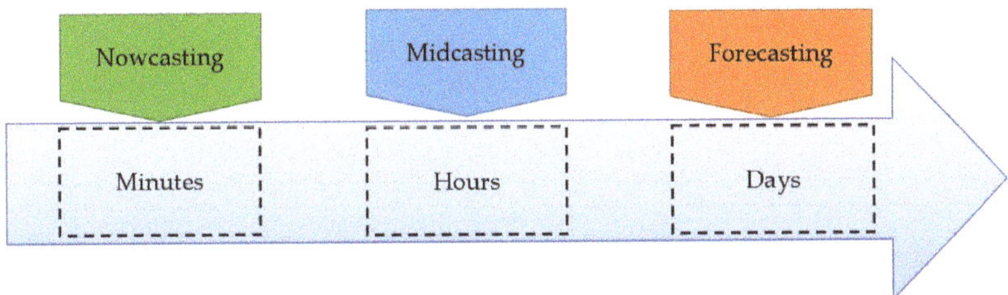

Figure 1. Prediction scale according to time horizon methods.

3.1. Time Horizon

The estimation of the output power of solar systems is necessary for the proper functioning of the electricity grid or for the optimal management of the energy flows that occur in the solar system. Before predicting the output of solar systems, it is essential to focus prediction on solar radiation. The prediction of components of solar radiation (global or diffuse) could be performed by several methods and the accuracy of a prediction model depends mainly on the time horizons [11].

For the development of any solar prediction model, we must contemplate the timescale (counted from a certain moment in which we make the prediction), which determines the future moment for which we make the predictions. The biggest drawback is its linear character, which makes it difficult for all problems to be properly modeled [22,24]. Models are very sensitive to unusual observations, forcing you to review the time series for detection and correcting before designing the prediction model [25].

3.1.1. Nowcasting Solar Radiation

The prediction horizon ranges from 15 min to a few hours, with no unanimity in its value [26]. In the short term, intra-hour forecasts are particularly useful for carrying out operations in the solar plant, balancing the grid, achieving automatic generation and trade control. Currently, it is very difficult to accurately predict solar radiation in the short term, as it involves knowing in advance how much energy solar plants produce instantly and this would be of great help in avoiding problems of supplying the line or avoiding surplus energy [27]. On this horizon, the statistical models that show the best performance are those that use satellite images with greater accuracy in nowcasting predictions [15,28] and statistical models for solar radiation time series, such as NWP [29–31]. Likewise, in recent years, several studies have been presented on the use of Machine Learning (ML) for solar radiation prediction using vector support machines (SVM) that show better performance in classification and regression analysis in time series [32,33].

3.1.2. Forecasting Solar Radiation

Long-term predictions: correspond to a horizon above 48 or 72 h, reaching a limit of 7 days. The larger the horizon, the greater the prediction errors, making it difficult to make reliable predictions of atmospheric variables above those 7 days [34–36]. Time horizons represent prediction analysis, as, as has been observed in the literature, ML models are able to demonstrate that they are the best alternative in time series data analysis. However, the need to reduce error correction has been modified to make way for ML regression model models using search optimization algorithms (SOA) that require accurate selection of their parameters to improve their performance [37,38].

4. Support Vector Machine Models

Machine Learning (ML) is the process of learning to convert experience into expertise. A concise explanation is that a computer program learns from the experience from data recorded in a period of time; the performance of the program is evaluated in time by improving experience [39]. In other words, ML models interpret patterns through learning based on data obtained in a period through a training process to make a prediction, generating new data that will measure the behavior of a phenomenon in the future [40–42].

According to the literature, there are two main specific techniques that ML uses to develop learning methods using information directly from data acquired [43]. There exist two main learning methods: supervised learning that uses from present to previous data to forecast events [44,45] and unsupervised learning that analyzes the patterns from non-classified data and deduces a function to describe the behavior of the system [46]. Additionally, there are other learning methods that derive from these principal methods: semi-supervised learning and multiple instance learning [47]. In this article, the supervised learning using a support vector machine (SVM) will be discussed. The Figure 2 summarizes the two principal methods and its most used models to predict solar radiation.

According to [8], supervised algorithms are currently the most used, and within this method are neural networks and vector support machines. These models have been widely used in recent years, with vector support machines being recently integrated to provide new solar prediction techniques. In this context, SVM belongs to a technique called supervised learning defined as a technique for identifying the behavior between an input and an output variable [48].

The theoretical basis of SVM is to minimize the structural risk related to the empirical risk from the training process and the confidence range from Vapnik Chervonenkis dimension (VC dimension) [49]. The complexity of the problem will be reduced when the VC dimension is smaller, making the risk smaller [50].

The architecture of an SVM is a versatile and configurable model based on a kernel machine that could be treated as a classification or regression problem according to Vapnik equations [51], while a support vector regression is only used for regression problems [11]. Therefore, an SVM is a great alternative to solve classification problems in forecasting time series [52–55].

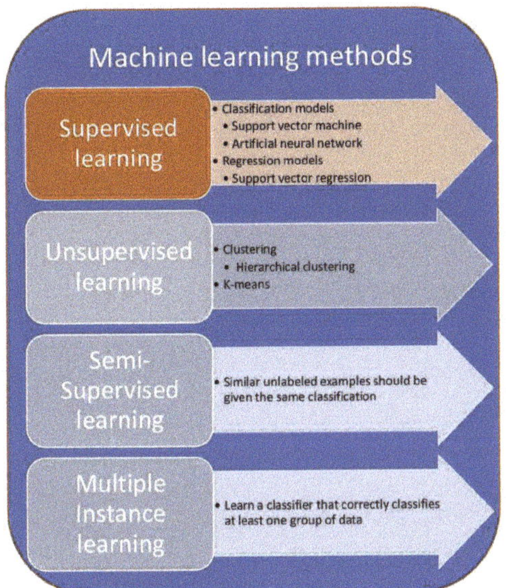

Figure 2. Machine learning techniques.

In order to predict a non-linear variable, such as solar radiation, it is necessary to have a data set x_i that represents the input data sample space, d_i which is the target value and to know the number of data points n [14]; therefore, it is possible to resort to Vapnik's equations theory. The input variables to SMV models are related to the objective variable (variable to predict), visualizing the mapped data in a non-linear function $f(x)$ [51]:

$$f(x) = \omega \cdot \phi(x) + b \tag{1}$$

where ω is the normal vector, b is a constant, also called bias term [56] and $\phi(x)$ is a large-dimensional spatial characteristic mapped by the space vector x. The coefficients ω and b are calculated by minimization using the following optimization problem [24]:

$$R_{svm}(f) = C\frac{1}{N}\sum_{i=1}^{N} x_{i=1} = L_e(f(x_i), y_i) + \frac{1}{2}\|w\|^2 \tag{2}$$

$$L_e(f(x_i), y_i) = \begin{cases} \text{si } |f(x), y| - \epsilon & \text{for } |f(x), y| \geq \epsilon \\ 0 & otherwise \end{cases} \tag{3}$$

where ϵ is a parameter of the model. $L_e(f(x_i), d_i)$ is the term that describes the ϵ-th missing function, which indicates that errors below epsilon are not penalized, d_i represents the solar radiation in the period i and $C\frac{1}{N}\sum_{i=1}^{N} L_\epsilon(f(x_i), d_i)$ defines the empirical error of the SVM model. $\frac{1}{2}\|w\|^2$ is the regularization term, C is the term that evaluates the error penalty function to regulate the compensation between the error or empirical risk and the term of regularization. The slack variables ς and ς^* indicate the excessive top and bottom skew, respectively. With these properties of the function to be optimized, it is possible to define Equation (2) as shown below [57]:

$$\text{minimize} \quad \frac{1}{2}\|w\|^2 + C\frac{1}{N}\sum_{i=1}^{N}(\varsigma_i + \varsigma_i^*) \tag{4}$$

$$\text{subject to:} = \begin{cases} |y_i - (\langle w, x_i \rangle + b) \geq \epsilon + \varsigma_i| \\ \langle w, x_i \rangle + b - y_i \leq \epsilon + \varsigma_i \\ \varsigma_i, \varsigma_i^* \geq 0 \end{cases} \tag{5}$$

To solve Equation (1), it is possible to use Lagrange and optimal constraints to obtain a non-linear regression function:

$$f(x) = \sum_{i=1}^{l} (\alpha_i - \alpha_i^*) K(x_i - x) + b \tag{6}$$

where α_i, α_i^* are Lagrange multipliers. The term $k(x_i - x)$ is defined as the kernel function [55]:

$$K(x_i - x) = \sum_{i=1}^{D} \phi_i(x) + \phi_i(y) \tag{7}$$

The general architecture form of a SVM is shown in Figure 3.

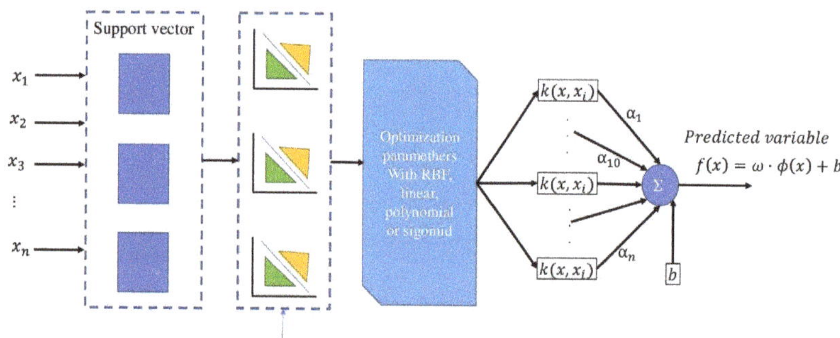

Figure 3. General architecture of a support vector maching (SVM) model according to [55].

4.1. Kernel Function Provided for SVM

SVM maps the data in a non-linear map to describe the linear process in the space of the predicted data according to the availability of data. This expression results in a simple linear combination problem attributed to the mapped space [58]. The kernel function allows a classification to be performed to form nonlinear boundaries to model complicated separating hyperplanes [59]. Figure 4 describes a projection from low to high dimension in space data.

The kernel parameters must be ideal to solve the problem classification according to the data to become separable in the next space. The four principal basic kernel functions are linear, polynomial, radial basis function (RBF) and sigmoid [57,60].

- Radial basis function (RBF): this function could perform nonlinear mapping of the samples into a higher dimensional feature space expressed by [61]:

$$K(x_i, x_j) = exp\left(-\frac{\|x_i - x_j\|^2 \sigma}{2}\right) \tag{8}$$

where σ is the kernel weight and x_i and x_j are the inputs to the i-th and j-th dimensions, respectively.

- Linear kernel function: According to [57], the linear function to obtain the SVM parameters is described by the following:

$$K(x_i, x_j) = x_i \cdot x_j \tag{9}$$

- Polynomial kernel function: this is a typical example of a global kernel, defined as follows:

$$K(x_i, x_j) = \left(x_i \cdot x_j\right)^q \tag{10}$$

where q is the degree of the polynom that will be used [62].
- Sigmoid kernel function: this function gives an explanation to practical viability [63]. This function is expressed as follows:

$$K(x_i, x_j) = tanh(v(x \cdot x_i) + c) \tag{11}$$

where v and c are adjustable kernel functions based on the data.

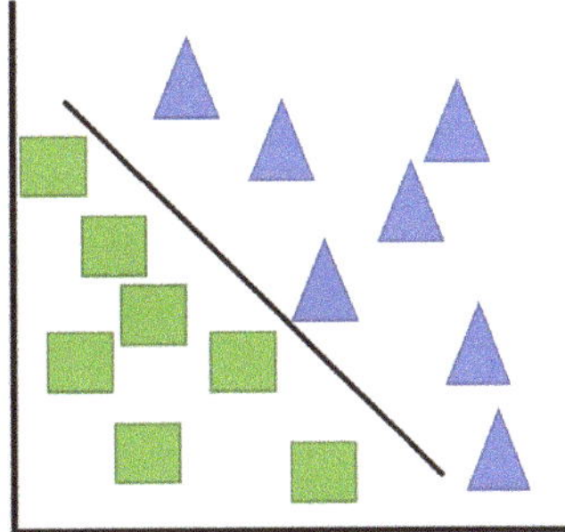

Input space   Feature space

Figure 4. Mapping of the kernel function.

4.2. Search Optimization Algorithms

In general, SVM models and ANN models are widely used in forecasting power consumption and solar radiation because of its performance and easy adaptation for nonlinear variables. However, it is important to improve new methods to find the best fitting parameters of the model [64].

The importance of using genetic algorithms as a solution to the Vapnik model function (Equation (1)) is to find the most optimal point to be able to obtain better prediction results. Intelligent search algorithms (SOA) are techniques for searching and exchanging information between the individuals of a population, which will be the ones that can solve a non-linear optimization problem [65]. SOA models are frequently used in forecasting methods where it is quite crucial to determine weights coefficients to reduce the error [66].

4.3. Performance Evaluation

It is necessary to know the performance of the SVM model. To evaluate the performance of forecasting solar radiation data, statistical indicators were used [67]. According to the literature, there are five popular indicators that could determine the accuracy of the predicted data [68,69]:

- Mean absolute percentage error (MAPE): It is used to express the absolute error of the predicted and observed variables in percentage [70]:

$$MAPE = \frac{1}{m}\sum_{i=1}^{m}\left|\frac{y_i - \hat{y}_i}{y_i}\right| \tag{12}$$

- Root mean square error ($RMSE$) sizes the goodness of the fit related to forecast with high errors [71]:

$$RMSE = \sqrt{\frac{1}{m}\sum_{i=1}^{m}(y_i - \hat{y}_i)^2} \quad (13)$$

- Mean bias error (MBE) indicates the deviation of predicted data from the observed data to provide the performance information of short and long-term use of the model [72]:

$$MBE = \frac{1}{m}\sum_{i=1}^{m}(y_i - \hat{y}_i) \quad (14)$$

- Mean absolute error (MAE) indicator gives a perspective of the performance of the prediction model by viewing how close the predicted variables are to observed variables [73]:

$$MAE = \frac{1}{m}\sum_{i=1}^{m}|y_i - \hat{y}_i| \quad (15)$$

- Relative root mean square error ($RRMSE$) quantifies the relative spread in the error [74,75]

$$RRMSE = \frac{1}{\bar{y}}\sqrt{\sum_{i=1}^{m}\frac{(y_i - \hat{y}_i)^2}{N}} \quad (16)$$

where y_i is the global solar radiation measured, $\hat{y}_i$ is the predicted global solar radiation, $\bar{y}$ is the mean global solar radiation, m is the number of forecast data points and N is the number of validation data. These indicators make it possible to know the efficiency of the SVM models: If the statistical indicator value is zero in the ideal case and presents a good performance if is closer to zero [8,57,67,76].

5. Results in Hybrid Techniques: Support Vector Machine with Search Algorithms Review

This work compiled and analyzed scientific articles focused on the construction of solar radiation prediction models with a hybrid method based on support vector machine and search optimization algoritms. The articles that are studied show the advances in this field and the solutions proposed to obtain predictive radiation data with the minimum error calculated with the statistical indicators.

SVMs relate to regularization networks and offers an advance on the ANN model. It is based on the theory of statistical learning that adopts least squares methods to solve the problem to least square solutions through a set of linear equations based on the minimization of structural risk.

Therefore, the SVM model can avoid excessive adjustment of the training data, does not require an iterative adjustment of the model parameters, has better generalization, requires few cores and has good performance [77]. To determine the appropriate choice of the prediction model with a support vector machine, the climatic variables of maximum and minimum air temperature (T_{max} and T_{min}, respectively), maximum and minimum relative humidity (H_{max} and H_{min}, respectively), wind speed (w_s), evaporation (E) and vapor pressure estimates (V) were the most used for future global solar radiation G_h [78–81]. This model can avoid the excessive adjustment of the training data, does not require an iterative adjustment of the model parameters, has better generalization, requires few cores and has good performance [55]. Figure 5 visualizes the general process to predict global solar radiation using the climatic variables.

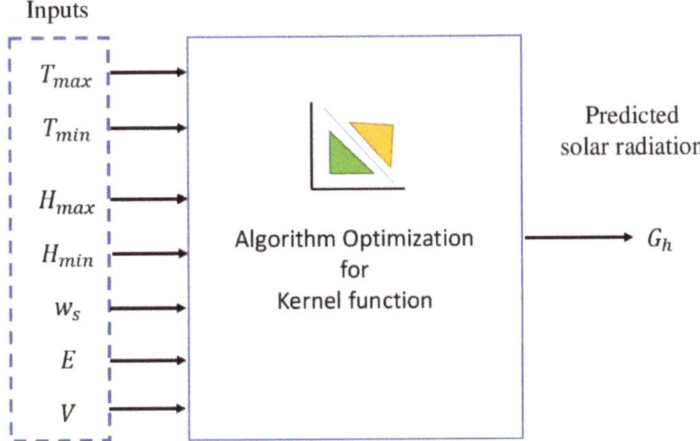

Figure 5. Training process using climatic variables for SVM models.

Different researchers have performed models based on SVM with no optimization models, and the results were prominent. Ref. [76] presented a hybrid SVM model in predicting solar radiation on two sites for PV panel surfaces using diffuse, direct and global solar radiation as input variables. The results show that the SVM model improves the prediction compared to the ANNs in the training phase with a maximum time of 0.0468 s, almost 5 times faster than an RNA model and 2 times faster in the testing phase, with 2.15 s. The stability of the model in this application lies in the use of these three radiation components with an RMSE three times smaller than an ANN, varying in a range from 18.34 to 31.15.

Authors from [36] proposed a solar prediction model based on SVM for one-hour ahead based on the typical climatic variables. They mentioned that SVM regression significantly improves the prediction accuracy according to the statistical indicators.

Authors from [67] proposed multiple SVM models using temperature equations with different kernel equations using only temperature variables from meteorological stations. The researchers observed that creating several empirical temperature-based equations and evaluating the models by several statistical indices, the prediction improves significantly in its performance.

In [82], a forecasting model based on SVM and using satellite images of clouds as the input space was presented. They used 4-year registers of cloud monitoring systems configured to perform the model by using large-scale data in multiple inputs and outputs. The performance SVM model was compared to other predicting models, providing a robust and great alternative to predict solar energy with $RMSE$ 10.86.

In [83], a model was developed to estimate global solar radiation based on a support vector machine (SVM) using an index that indicates air quality as an input variable to evaluate the performance of this technique. Compared to existing models, such as neural networks, the model presents a great performance in the accuracy of global solar radiation models by using an air quality indicator as an additional input parameter. The related works only used SVM models. Table 1 summarizes the findings about using SVM as an important alternative to ANN or other models to predict solar radiation.

Table 1. Techniques of solar radiation prediction using SVM.

Reference	Analysis of Results	Time Horizon	Kernel Function	MAPE	RMSE	MBE	MAE	RRMSE
[76]	Compared ANN and SVM in predicting the solar radiation	1 day	polynomial	-	28.39 $\frac{W}{m^2}$	-	-	-
[83]	SVM performs better than other models if air quality index is used in the models using polynomial kernel function	1 h	Polynomial	8.24%	-	-	-	-
[36]	They compared the SVM with (ANN) and Non linear autoregressive (NAR), showing that SVM performs well in prediction accuracy	1 h	RBF	-	4.26 $\frac{W}{m^2}$	-	-	-
[82]	SVM performs great than ANN models and can be effectively used for grid operations and energy management systems	15 min	RBF	-	28.00 $\frac{W}{m^2}$	-	-	-
[67]	Polynomial kernel function performs great using temperature equations in SVM models	1 h	Polynomial	-	0.83 $\frac{MJ}{m^2}$	-	-	9.00%

In recent years, the state-of-the-art with respect to SVM has been demanding due to its high performance, so various methods of improving these models have been developed, such as integrating a search algorithm to the SVM model. Some works have proposed various hybrid techniques based on the flow diagram in the Figure 6.

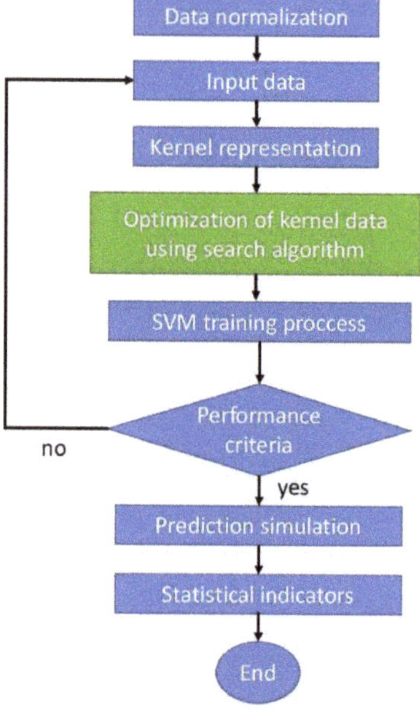

Figure 6. Flow chart of SVM model with search algorithm.

The work of [14] showed an SVM with the Firefly Algorithm (FFA) to estimate future solar irradiance. A Firefly Algorithm (FFA) was used to calculate the best kernel parameters of the SVM model to obtain the most accurate prediction data. The model presents a better performance compared to ANN, which has RMSE = 1.86% and a MAPE value of 11.51%. The results are promising for the use of FFA with SVM as a great alternative to predicting the global component of the solar radiation. Additionally, they specified a risk in zones with no constant climatic conditions.

In [84], a new approach was developed using data mining techniques to model and improve the prediction of hourly global solar radiation the next day. They report an RMSE value of 23.5% when the independent variables were the values of the meteorological parameters of the previous day and an RMSE of 22.9% when the independent variables were the forecasts of meteorological variables for the same day to be forecast.

The authors [85] developed an SVM model optimized by glowworm swarm optimization (GSO) to estimate solar radiation. Employing the Eclat algorithm (data mining) to choose the appropriate predictors, the support vector machine (SVM) with penalty function of the model structure and the GSO algorithm to improve forecast accuracy, selecting the appropriate parameters. The tests were carried out in four areas of the USA using the input variables of average daily airmass, dew point, relative humidity, opaque cloud cover, wind speed, vertical wind shear, pressure, albedo, zenith angle, azimuth angle, net radiation, global normal radiation and global extraterrestrial radiation. The results show a great performance—RMSE = 0.43 $\frac{W}{m^2}$ and MAPE 5.64%.

In [86], an SVM-FFA model for the prediction of the monthly mean daily horizontal global solar radiation in the port of Bandar Abbass was proposed, located in the southern coastal region of Iran. They also show that their model is highly efficient in estimating the monthly mean daily horizontal global solar radiation. With the following statistical indicators, MAPE = 3.3252.

Ref. [87] presented an SVM model with optimization using GA to compare its performance with a grid search evaluating the accuracy by the analysis of the parameters of the four kernel functions. They found that the grid search could be a good alternative only if a low dimensional dataset, which is only present in few cases of classifications problems, but GA is the best alternative in more cases, presenting stability above 15.9-times that of a grid search.

Research from [57] presented a study of three models: adaptive neuro-fuzzy inference system (ANFIS), ANN and (SVM) models using GA in the three cases to estimate the kernel model parameters in order to compare their accuracy and performance in daily global solar radiation. The results show a huge advantage using SVM-GA to predict solar radiation. They found that SVM-GA also determined the best kernel parameters by evaluating the error to the global minimum convergence.

In [55], a wavelet-coupled vector support machine model for predicting global solar incidents using a limited meteorological data set for the city of Brisbane, Australia was performed. The data were decomposed into a subset of wavelets, transforming the input space into discrete variables to create new time series using the Daubechies-2 wavelet to a detailed level. This hybrid model obtained an approximation of $R = 0.965$. The input variables used in this model are the basic climatic variables and add hours of sunshine St to predict the daily global incident solar radiation (R_n). They also mentioned that climatic anomalies could make the W-SVM model less accurate for solar radiation forecasting.

The authors from [88] presented a GA-SVM model to predict the short-term power forecasting of a PV system on a residential scale. GA was used to find the optimal parameters values for a kernel within a base classifier of the SVM. They used climatic variables and added the solar radiation to predict the power load of a photovoltaic system, that shows that SVM is multi-configurable according to the problem. The results show that GA-SVM outperforms the conventional SVM. The GA shows its great performance by finding a local minimum that is translated in a high-accuracy solar radiation forecasting. Table 2 summarizes the findings on the use of SVM optimized with search algorithm techniques.

Ref. [89] performed a comparison between SVM models and copula-based nonlinear quaintly regression (CNQR) to predict dayli diffuse solar radiation and mentioned the principal risk working with SVM models: There is a possibility that SVM could not perform an accurate prediction if lower parameters are considered in the space data; with more input parameters, the accuracy will increase but the parameter optimization increases considerable, leading to a high cost for the prediction.

The authors from [90], developed three hybrid models to predict the daily solar radiation in urban zones: SVM-PSO model, Bat algorithm with SVM and Whale optimization algorithm with SVM using the general climatic data and adding the Ozone (O_3) in the space variables. They found that the O_3 variable improves the accuracy of the predicted data compared to other parameters such as air pollution, using the RMSD statistical indicator to measure the performance, with values of 11.1%, 10.0% and 10.4%, respectively, for the three developed models.

Table 2. Techniques of solar radiation prediction using SVM and search algorithms.

Reference	Analysis of Results	Time Horizon	Optimization Model	Kernel Function	MAPE	RMSE	MAE	RRMSE
[14]	SVM-FFA present a better performance in comparisson with ANN models	1 h	FFA	RBF	11.51%	-	-	1.86%
[84]	Mining data to forecasting hourly global solar radiation	1 h	SVM-R	Sigmoid	-	119 $\frac{W}{m^2}$	79 $\frac{W}{m^2}$	22.90%
[85]	Evaluated the performances of SVM, HARD-RIDGE-SVM, SVM-HARD and GSO-SVM-HARD model	1 day	GSO	Hilbert space	5.64%	0.43 $\frac{W}{m^2}$	-	-
[86]	SVM-FFA present a better performance compared to the ANN, GP, and ARMA techniques	1 month	FFA	RBF	3.32%	0.18 $\frac{kW}{m^2}$	-	3.73%
[87]	SVM parameter optimization using GA is more than 15.9 times faster than using grid search. F-measure was used to evaluate the performance	-	GA	RBF sigmoid	8.24%	-	-	-
[55]	SVM incorporates a discrete wavelet transformation algorithm for pre-processing of inputs	1 day	Wavelet	RBF	4.69%	1.18 $\frac{MJ}{m^2}$	0.92 $\frac{MJ}{m^2}$	5.94%
[57]	SVM-GA models has higher prediction accuracy in tropical warm sub-humidthan ANN model	10 min	GA	-	-	2.57 $\frac{MJ}{m^2}$	1.97 $\frac{MJ}{m^2}$	-
[88]	GA-SVM outperform SVM classical models in classification climatic data	1 h	GA	Gaussian	1.70%	11.22 $\frac{W}{m^2}$	-	-
[89]	SVM-FFA better than copula-base nonlinear quantile regression	1 day	-	RBF	-	1.18 $\frac{MJ}{m^2}$	-	18.00%
[90]	Mentioned that SVM with PSO convergence to local optimal solution faster than the other proposal algorithms.	1 h	PSO	RBF	-	-	0.99 $\frac{MJ}{m^2}$	2.90%

6. Discussion and Conclusions

This article presented a state-of-the-art in solar radiation prediction techniques through the use of support vector machine models (SVM) and the optimization of the search for the parameters that guarantee a favorable performance in the accuracy of the forecasting data.

It was observed that SVMs by themselves show a better performance in predicting solar radiation than artificial neural networks and other prediction models such as auto regression. In this context, it is possible to define these models as premature techniques, which today are consolidated as the best predictive models, since these models emerged 6 years ago [8]. The SVM models show an improvement when evaluating the polynomial kernel functions using only as a basis the climatic variables of maximum and minimum air temperature (T_{max} and T_{min}, respectively), maximum and minimum relative humidity (H_{max} and H_{min}, respectively), wind speed (w_s), evaporation (E) and vapor pressure estimates (V). That base of the input space is sufficient to perform a prediction model based on SVM. However, it is possible that the polynomial function demands a high cost in performance because it can increase in degree, so another alternative is the use of a radial basis function (RBF). According to the literature, RBF to estimate the kernel parameters can improve the accuracy up to 7% compared to higher degree polynomial functions. Likewise, the most widely used validation method is $RMSE$, since the observed error contributes to the tuning of the weights of the SVM model. Additionally, according to [76], SVM could be 5 times faster than ANN in the training phase and 2 times faster in the testing phase.

The execution time of each optimization algorithm will depend on the number of input variables to the prediction model, the number of data and the prediction horizon. In order to obtain a more robust model, it must be calibrated with the variables and geographical conditions, since they change according to the area, so the accuracy totally depends on the location. These search algorithms considerably revolutionize the performance of the SVM model compared to a simple SVM model, contributing to obtaining a lower prediction error in the output of the solar radiation data. In addition, in these models, the RBF kernel function presents better performance in the search for parameters due to the inclusion of these search algorithms. These algorithms can be evaluated based on MAPE, RMSE and MAE as the main statistical indicators to evaluate the predicted data of solar radiation that measure the performance of the model.

Finally, the versatility of the model using only climate data for the prediction of solar systems is noteworthy, such that in the future they may be a beneficial alternative in the sizing, generation and management of alternative energy to contribute to the energy transition that in the coming years will be one of the main issues that researchers will tackle to solve problems.

Author Contributions: Conceptualization, J.M.Á.-A. and J.G.R.-M.; investigation, J.M.Á.-A.; validation, G.R.-L., E.V.-R.J. and S.A.O.-B.; writing—original draft preparation, J.M.Á.-A.; discussion, J.G.R.-M.; writing—review and editing, M.T.-P. and J.G.R.-M.; formal analysis, J.G.R.-M. All authors have read and agreed to the published version of the manuscript.

Funding: This research received no external funding.

Institutional Review Board Statement: Not applicable.

Informed Consent Statement: Not applicable.

Data Availability Statement: The data presented in this study are available on request from the corresponding author.

Conflicts of Interest: The authors declare no conflict of interest.

Abbreviations

The following abbreviations are used in this manuscript:

NWP	Numerical weather prediction
ML	Machine learning
SVM	Support vector machine
SOA	Search optimization algorithms
G_h	Global solar radiation
$f(x)$	Vapnik Function
x	Input data
ω	Normal vector
b	Bias term
$\phi(x)$	Large-dimensional spatial characteristic
(C, ϵ)	parameters of the model
ς	Excessive top skew
ς^*	Excessive buttom skew
K	Kernel
α	Lagrange multiplicators
i	index
$\frac{1}{2}\|w\|^2$	Regularization term
RBF	Radial basis function
q	Degree of polynom
GA	Genetic Algorithm
PSO	Particle Swarn Optimization
$MAPE$	Coefficient of determination
$RMSE$	Root mean square error
MBE	Mean bias error
MAE	Mean absolute error
$RRMSE$	Relative root mean square error
T_{max}	Maximum temperature
T_{min}	Minimum temperature
H_{max}	Maximum relative humidity
H_{min}	Minimum relative humidity
w_{max}	Wind speed
E_{min}	Evaporation
V	Vapor pressure

References

1. Cradden, L.; Burnett, D.; Agarwal, A.; Harrison, G. Climate change impacts on renewable electricity generation. *Infrastruct. Asset Manag.* **2015**, *2*, 131–142. [CrossRef]
2. Coady, D.; Parry, I.; Sears, L.; Shang, B. How large are global fossil fuel subsidies? *World Dev.* **2017**, *91*, 11–27. [CrossRef]
3. Gustavsson, L.; Haus, S.; Lundblad, M.; Lundström, A.; Ortiz, C.A.; Sathre, R.; Le Truong, N.; Wikberg, P.E. Climate change effects of forestry and substitution of carbon-intensive materials and fossil fuels. *Renew. Sustain. Energy Rev.* **2017**, *67*, 612–624. [CrossRef]
4. Gustavsson, L.; Haus, S.; Ortiz, C.A.; Sathre, R.; Le Truong, N. Climate effects of bioenergy from forest residues in comparison to fossil energy. *Appl. Energy* **2015**, *138*, 36–50. [CrossRef]
5. Cecati, C.; Ciancetta, F.; Siano, P. A multilevel inverter for photovoltaic systems with fuzzy logic control. *IEEE Trans. Ind. Electron.* **2010**, *57*, 4115–4125. [CrossRef]
6. Chang, T.P.; Liu, F.J.; Ko, H.H.; Huang, M.C. Oscillation characteristic study of wind speed, global solar radiation and air temperature using wavelet analysis. *Appl. Energy* **2017**, *190*, 650–657. [CrossRef]
7. Álvarez-Alvarado, J.M.; Ríos-Moreno, J.G.; Ventura-Ramos, E.J.; Ronquillo-Lomeli, G.; Trejo-Perea, M. An alternative methodology to evaluate sites using climatology criteria for hosting wind, solar, and hybrid plants. *Energy Sources Part A Recover. Util. Environ. Eff.* **2020**, 1–18. [CrossRef]
8. Voyant, C.; Notton, G.; Kalogirou, S.; Nivet, M.L.; Paoli, C.; Motte, F.; Fouilloy, A. Machine learning methods for solar radiation forecasting: A review. *Renew. Energy* **2017**, *105*, 569–582. [CrossRef]
9. Suganthi, L.; Samuel, A.A. Energy models for demand forecasting—A review. *Renew. Sustain. Energy Rev.* **2012**, *16*, 1223–1240. [CrossRef]

10. Aler, R.; Galván, I.M.; Ruiz-Arias, J.A.; Gueymard, C.A. Improving the separation of direct and diffuse solar radiation components using machine learning by gradient boosting. *Sol. Energy* **2017**, *150*, 558–569. [CrossRef]
11. Lauret, P.; Voyant, C.; Soubdhan, T.; David, M.; Poggi, P. A benchmarking of machine learning techniques for solar radiation forecasting in an insular context. *Sol. Energy* **2015**, *112*, 446–457. [CrossRef]
12. Xue, X. Prediction of daily diffuse solar radiation using artificial neural networks. *Int. J. Hydrog. Energy* **2017**, *42*, 28214–28221. [CrossRef]
13. Renno, C.; Petito, F.; Gatto, A. ANN model for predicting the direct normal irradiance and the global radiation for a solar application to a residential building. *J. Clean. Prod.* **2016**, *135*, 1298–1316. [CrossRef]
14. Olatomiwa, L.; Mekhilef, S.; Shamshirband, S.; Mohammadi, K.; Petković, D.; Sudheer, C. A support vector machine–firefly algorithm-based model for global solar radiation prediction. *Sol. Energy* **2015**, *115*, 632–644. [CrossRef]
15. Zhang, J.; Verschae, R.; Nobuhara, S.; Lalonde, J.F. Deep photovoltaic nowcasting. *Sol. Energy* **2018**, *176*, 267–276. [CrossRef]
16. Rabehi, A.; Guermoui, M.; Lalmi, D. Hybrid models for global solar radiation prediction: A case study. *Int. J. Ambient. Energy* **2020**, *41*, 31–40. [CrossRef]
17. Elminir, H.K.; Areed, F.F.; Elsayed, T.S. Estimation of solar radiation components incident on Helwan site using neural networks. *Sol. Energy* **2005**, *79*, 270–279. [CrossRef]
18. Gueymard, C.A. The sun's total and spectral irradiance for solar energy applications and solar radiation models. *Sol. Energy* **2004**, *76*, 423–453. [CrossRef]
19. Ulgen, K.; Hepbasli, A. Diffuse solar radiation estimation models for Turkey's big cities. *Energy Convers. Manag.* **2009**, *50*, 149–156. [CrossRef]
20. Jiang, Y. Daily diffuse solar radiation at Beijing. In Proceedings of the IEEE 2009 World Non-Grid-Connected Wind Power and Energy Conference, Nanjing, China, 24–26 September 2009; pp. 1–4.
21. Behar, O.; Khellaf, A.; Mohammedi, K. Comparison of solar radiation models and their validation under Algerian climate—The case of direct irradiance. *Energy Convers. Manag.* **2015**, *98*, 236–251. [CrossRef]
22. Arbizu-Barrena, C.; Ruiz-Arias, J.A.; Rodríguez-Benítez, F.J.; Pozo-Vázquez, D.; Tovar-Pescador, J. Short-term solar radiation forecasting by advecting and diffusing MSG cloud index. *Sol. Energy* **2017**, *155*, 1092–1103. [CrossRef]
23. Ghimire, S.; Deo, R.C.; Downs, N.J.; Raj, N. Global solar radiation prediction by ANN integrated with European Centre for medium range weather forecast fields in solar rich cities of Queensland Australia. *J. Clean. Prod.* **2019**, *216*, 288–310. [CrossRef]
24. Zeng, J.; Qiao, W. Short-term solar power prediction using a support vector machine. *Renew. Energy* **2013**, *52*, 118–127. [CrossRef]
25. Zamarbide Ducun, I. Predicción de Radiacion Solar a Corto y Medio Plazo. Master's Thesis, Universidad Publica de Navarra, Pamplona, Spain, 2014.
26. Sanfilippo, A.; Martin-Pomares, L.; Mohandes, N.; Perez-Astudillo, D.; Bachour, D. An adaptive multi-modeling approach to solar nowcasting. *Sol. Energy* **2016**, *125*, 77–85. [CrossRef]
27. Paulescu, M.; Paulescu, E. Short-term forecasting of solar irradiance. *Renew. Energy* **2019**, *143*, 985–994. [CrossRef]
28. Gallucci, D.; Romano, F.; Cersosimo, A.; Cimini, D.; Di Paola, F.; Gentile, S.; Geraldi, E.; Larosa, S.; Nilo, S.T.; Ricciardelli, E.; et al. Nowcasting surface solar irradiance with AMESIS via motion vector fields of MSG-SEVIRI data. *Remote Sens.* **2018**, *10*, 845. [CrossRef]
29. Lima, F.J.; Martins, F.R.; Pereira, E.B.; Lorenz, E.; Heinemann, D. Forecast for surface solar irradiance at the Brazilian Northeastern region using NWP model and artificial neural networks. *Renew. Energy* **2016**, *87*, 807–818. [CrossRef]
30. Lorenz, E.; Kühnert, J.; Heinemann, D. Short term forecasting of solar irradiance by combining satellite data and numerical weather predictions. In Proceedings of the 27th European PV Solar Energy Conference (EU PVSEC), Frankfurt, Germany, 24–28 September 2012; Volume 2428, p. 44014405.
31. Tiwari, S.; Sabzehgar, R.; Rasouli, M. Short term solar irradiance forecast using numerical weather prediction (nwp) with gradient boost regression. In Proceedings of the 2018 9th IEEE International Symposium on Power Electronics for Distributed Generation Systems (PEDG), Charlotte, NC, USA, 25–28 June 2018; pp. 1–8.
32. Sun, X.; Zhang, T. Solar power prediction in smart grid based on nwp data and an improved boosting method. In Proceedings of the 2017 IEEE International Conference on Energy Internet (ICEI), Beijing, China, 17–21 April 2017; pp. 89–94.
33. Eseye, A.T.; Zhang, J.; Zheng, D. Short-term photovoltaic solar power forecasting using a hybrid Wavelet-PSO-SVM model based on SCADA and Meteorological information. *Renew. Energy* **2018**, *118*, 357–367. [CrossRef]
34. Notton, G.; Voyant, C.; Fouilloy, A.; Duchaud, J.L.; Nivet, M.L. Some applications of ANN to solar radiation estimation and forecasting for energy applications. *Appl. Sci.* **2019**, *9*, 209. [CrossRef]
35. Yesilbudak, M.; Colak, M.; Bayindir, R. What are the Current Status and Future Prospects in Solar Irradiance and Solar Power Forecasting? *Int. J. Renew. Energy Res.* **2018**, *8*, 635–648.
36. Bae, K.Y.; Jang, H.S.; Sung, D.K. Hourly solar irradiance prediction based on support vector machine and its error analysis. *IEEE Trans. Power Syst.* **2016**, *32*, 935–945. [CrossRef]
37. Zendehboudi, A.; Baseer, M.; Saidur, R. Application of support vector machine models for forecasting solar and wind energy resources: A review. *J. Clean. Prod.* **2018**, *199*, 272–285. [CrossRef]
38. Li, J.; Ward, J.K.; Tong, J.; Collins, L.; Platt, G. Machine learning for solar irradiance forecasting of photovoltaic system. *Renew. Energy* **2016**, *90*, 542–553. [CrossRef]

39. Shalev-Shwartz, S.; Ben-David, S. *Understanding Machine Learning: From Theory to Algorithms*; Cambridge University Press: Cambridge, UK, 2014.
40. Barboza, F.; Kimura, H.; Altman, E. Machine learning models and bankruptcy prediction. *Expert Syst. Appl.* **2017**, *83*, 405–417. [CrossRef]
41. Christodoulou, E.; Ma, J.; Collins, G.S.; Steyerberg, E.W.; Verbakel, J.Y.; Van Calster, B. A systematic review shows no performance benefit of machine learning over logistic regression for clinical prediction models. *J. Clin. Epidemiol.* **2019**, *110*, 12–22. [CrossRef] [PubMed]
42. Mao, X.; Yang, H.; Huang, S.; Liu, Y.; Li, R. Extractive summarization using supervised and unsupervised learning. *Expert Syst. Appl.* **2019**, *133*, 173–181. [CrossRef]
43. Bao, W.; Lianju, N.; Yue, K. Integration of unsupervised and supervised machine learning algorithms for credit risk assessment. *Expert Syst. Appl.* **2019**, *128*, 301–315. [CrossRef]
44. Shami, M.; Verhelst, W. An evaluation of the robustness of existing supervised machine learning approaches to the classification of emotions in speech. *Speech Commun.* **2007**, *49*, 201–212. [CrossRef]
45. Jiang, T.; Gradus, J.L.; Rosellini, A.J. Supervised machine learning: A brief primer. *Behav. Ther.* **2020**, *51*, 675–687. [CrossRef]
46. Saravanan, R.; Sujatha, P. A state of art techniques on machine learning algorithms: A perspective of supervised learning approaches in data classification. In Proceedings of the IEEE 2018 Second International Conference on Intelligent Computing and Control Systems (ICICCS), Madurai, India, 14–15 June 2018; pp. 945–949.
47. Komura, D.; Ishikawa, S. Machine learning methods for histopathological image analysis. *Comput. Struct. Biotechnol. J.* **2018**, *16*, 34–42. [CrossRef]
48. Yadav, A.K.; Chandel, S. Solar radiation prediction using Artificial Neural Network techniques: A review. *Renew. Sustain. Energy Rev.* **2014**, *33*, 772–781. [CrossRef]
49. Fan, J.; Wang, X.; Wu, L.; Zhou, H.; Zhang, F.; Yu, X.; Lu, X.; Xiang, Y. Comparison of Support Vector Machine and Extreme Gradient Boosting for predicting daily global solar radiation using temperature and precipitation in humid subtropical climates: A case study in China. *Energy Convers. Manag.* **2018**, *164*, 102–111. [CrossRef]
50. Kazem, H.A.; Yousif, J.H.; Chaichan, M.T. Modeling of daily solar energy system prediction using support vector machine for Oman. *Int. J. Appl. Eng. Res.* **2016**, *11*, 10166–10172.
51. Vapnik, V.; Izmailov, R. Knowledge transfer in SVM and neural networks. *Ann. Math. Artif. Intell.* **2017**, *81*, 3–19. [CrossRef]
52. Yang, X.; Jiang, F.; Liu, H. Short-term solar radiation prediction based on SVM with similar data. In Proceedings of the 2nd IET Renewable Power Generation Conference (RPG 2013), Beijing, China, 9–11 September 2013.
53. Belaid, S.; Mellit, A. Prediction of daily and mean monthly global solar radiation using support vector machine in an arid climate. *Energy Convers. Manag.* **2016**, *118*, 105–118. [CrossRef]
54. Ibrahim, I.A.; Khatib, T. A novel hybrid model for hourly global solar radiation prediction using random forests technique and firefly algorithm. *Energy Convers. Manag.* **2017**, *138*, 413–425. [CrossRef]
55. Deo, R.C.; Wen, X.; Qi, F. A wavelet-coupled support vector machine model for forecasting global incident solar radiation using limited meteorological dataset. *Appl. Energy* **2016**, *168*, 568–593. [CrossRef]
56. Wang, H.; Hu, D. Comparison of SVM and LS-SVM for regression. In Proceedings of the IEEE 2005 International Conference on Neural Networks and Brain, Beijing, China, 13–15 October 2005; Volume 1, pp. 279–283.
57. Quej, V.H.; Almorox, J.; Arnaldo, J.A.; Saito, L. ANFIS, SVM and ANN soft-computing techniques to estimate daily global solar radiation in a warm sub-humid environment. *J. Atmos. Sol. Terr. Phys.* **2017**, *155*, 62–70. [CrossRef]
58. Ramedani, Z.; Omid, M.; Keyhani, A.; Khoshnevisan, B.; Saboohi, H. A comparative study between fuzzy linear regression and support vector regression for global solar radiation prediction in Iran. *Sol. Energy* **2014**, *109*, 135–143. [CrossRef]
59. Prajapati, G.L.; Patle, A. On performing classification using SVM with radial basis and polynomial kernel functions. In Proceedings of the IEEE 2010 3rd International Conference on Emerging Trends in Engineering and Technology, Goa, India, 19–21 November 2010; pp. 512–515.
60. Hassan, M.Z.; Ali, M.E.K.; Ali, A.S.; Kumar, J. Forecasting day-ahead solar radiation using machine learning approach. In Proceedings of the IEEE 2017 4th Asia-Pacific World Congress on Computer Science and Engineering (APWC on CSE), Nadi, Fiji, 11–13 December 2017; pp. 252–258.
61. Zhen, Z.; Wang, F.; Sun, Y.; Mi, Z.; Liu, C.; Wang, B.; Lu, J. SVM based cloud classification model using total sky images for PV power forecasting. In Proceedings of the 2015 IEEE Power & Energy Society Innovative Smart Grid Technologies Conference (ISGT), Washington, DC, USA, 18–20 February 2015; pp. 1–5.
62. Smits, G.F.; Jordaan, E.M. Improved SVM regression using mixtures of kernels. In Proceedings of the IEEE 2002 International Joint Conference on Neural Networks. IJCNN'02 (Cat. No. 02CH37290), Honolulu, HI, USA, 12–17 May 2002; Volume 3, pp. 2785–2790.
63. Lin, H.T.; Lin, C.J. A study on sigmoid kernels for SVM and the training of non-PSD kernels by SMO-type methods. *Submitt. Neural Comput.* **2003**, *3*, 16.
64. Wu, Y.H.; Shen, H. Grey-related least squares support vector machine optimization model and its application in predicting natural gas consumption demand. *J. Comput. Appl. Math.* **2018**, *338*, 212–220. [CrossRef]
65. Chavero-Navarrete, E.; Trejo-Perea, M.; Jáuregui-Correa, J.C.; Carrillo-Serrano, R.V.; Ríos-Moreno, J.G. Expert control systems for maximum power point tracking in a wind turbine with PMSG: State of the art. *Appl. Sci.* **2019**, *9*, 2469. [CrossRef]

66. Wang, J.; Jiang, H.; Wu, Y.; Dong, Y. Forecasting solar radiation using an optimized hybrid model by Cuckoo Search algorithm. *Energy* **2015**, *81*, 627–644. [CrossRef]
67. Chen, J.L.; Liu, H.B.; Wu, W.; Xie, D.T. Estimation of monthly solar radiation from measured temperatures using support vector machines—A case study. *Renew. Energy* **2011**, *36*, 413–420. [CrossRef]
68. Jiang, H.; Dong, Y. Global horizontal radiation forecast using forward regression on a quadratic kernel support vector machine: Case study of the Tibet Autonomous Region in China. *Energy* **2017**, *133*, 270–283. [CrossRef]
69. Wang, F.; Zhen, Z.; Wang, B.; Mi, Z. Comparative study on KNN and SVM based weather classification models for day ahead short term solar PV power forecasting. *Appl. Sci.* **2018**, *8*, 28. [CrossRef]
70. Chen, S.; Gooi, H.; Wang, M. Solar radiation forecast based on fuzzy logic and neural networks. *Renew. Energy* **2013**, *60*, 195–201. [CrossRef]
71. Verbois, H.; Huva, R.; Rusydi, A.; Walsh, W. Solar irradiance forecasting in the tropics using numerical weather prediction and statistical learning. *Sol. Energy* **2018**, *162*, 265–277. [CrossRef]
72. Benatiallah, D.; Bouchouicha, K.; Benatiallah, A.; Harrouz, A.; Nasri, B. Forecasting of solar radiation using an empirical model. *Alger. J. Renew. Energy Sustain. Dev.* **2019**, *1*, 212–219. [CrossRef]
73. Li, Z.; Rahman, S.; Vega, R.; Dong, B. A hierarchical approach using machine learning methods in solar photovoltaic energy production forecasting. *Energies* **2016**, *9*, 55. [CrossRef]
74. Chen, H.; Wu, W.; Liu, H.B. Assessing the relative importance of climate variables to rice yield variation using support vector machines. *Theor. Appl. Climatol.* **2016**, *126*, 105–111. [CrossRef]
75. Chu, Y.; Pedro, H.T.; Kaur, A.; Kleissl, J.; Coimbra, C.F. Net load forecasts for solar-integrated operational grid feeders. *Solar Energy* **2017**, *158*, 236–246. [CrossRef]
76. Ramli, M.A.; Twaha, S.; Al-Turki, Y.A. Investigating the performance of support vector machine and artificial neural networks in predicting solar radiation on a tilted surface: Saudi Arabia case study. *Energy Convers. Manag.* **2015**, *105*, 442–452. [CrossRef]
77. Salcedo-Sanz, S.; Casanova-Mateo, C.; Pastor-Sánchez, A.; Sánchez-Girón, M. Daily global solar radiation prediction based on a hybrid Coral Reefs Optimization–Extreme Learning Machine approach. *Sol. Energy* **2014**, *105*, 91–98. [CrossRef]
78. Sharafati, A.; Khosravi, K.; Khosravinia, P.; Ahmed, K.; Salman, S.A.; Yaseen, Z.M.; Shahid, S. The potential of novel data mining models for global solar radiation prediction. *Int. J. Environ. Sci. Technol.* **2019**, *16*, 7147–7164. [CrossRef]
79. Hou, M.; Zhang, T.; Weng, F.; Ali, M.; Al-Ansari, N.; Yaseen, Z.M. Global solar radiation prediction using hybrid online sequential extreme learning machine model. *Energies* **2018**, *11*, 3415. [CrossRef]
80. Kisi, O.; Heddam, S.; Yaseen, Z.M. The implementation of univariable scheme-based air temperature for solar radiation prediction: New development of dynamic evolving neural-fuzzy inference system model. *Appl. Energy* **2019**, *241*, 184–195. [CrossRef]
81. Citakoglu, H.; Babayigit, B.; Haktanir, N.A. Solar radiation prediction using multi-gene genetic programming approach. *Theor. Appl. Climatol.* **2020**, *142*, 885–897. [CrossRef]
82. Jang, H.S.; Bae, K.Y.; Park, H.S.; Sung, D.K. Solar power prediction based on satellite images and support vector machine. *IEEE Trans. Sustain. Energy* **2016**, *7*, 1255–1263. [CrossRef]
83. Yao, W.; Zhang, C.; Hao, H.; Wang, X.; Li, X. A support vector machine approach to estimate global solar radiation with the influence of fog and haze. *Renew. Energy* **2018**, *128*, 155–162. [CrossRef]
84. Jiménez-Pérez, P.F.; Mora-López, L. Modeling and forecasting hourly global solar radiation using clustering and classification techniques. *Sol. Energy* **2016**, *135*, 682–691. [CrossRef]
85. Jiang, H.; Dong, Y. A nonlinear support vector machine model with hard penalty function based on glowworm swarm optimization for forecasting daily global solar radiation. *Energy Convers. Manag.* **2016**, *126*, 991–1002. [CrossRef]
86. Shamshirband, S.; Mohammadi, K.; Tong, C.W.; Zamani, M.; Motamedi, S.; Ch, S. A hybrid SVM-FFA method for prediction of monthly mean global solar radiation. *Theor. Appl. Climatol.* **2016**, *125*, 53–65. [CrossRef]
87. Syarif, I.; Prugel-Bennett, A.; Wills, G. SVM parameter optimization using grid search and genetic algorithm to improve classification performance. *Telkomnika* **2016**, *14*, 1502. [CrossRef]
88. VanDeventer, W.; Jamei, E.; Thirunavukkarasu, G.S.; Seyedmahmoudian, M.; Soon, T.K.; Horan, B.; Mekhilef, S.; Stojcevski, A. Short-term PV power forecasting using hybrid GASVM technique. *Renew. Energy* **2019**, *140*, 367–379. [CrossRef]
89. Liu, Y.; Zhou, Y.; Chen, Y.; Wang, D.; Wang, Y.; Zhu, Y. Comparison of support vector machine and copula-based nonlinear quantile regression for estimating the daily diffuse solar radiation: A case study in China. *Renew. Energy* **2020**, *146*, 1101–1112. [CrossRef]
90. Fan, J.; Wu, L.; Ma, X.; Zhou, H.; Zhang, F. Hybrid support vector machines with heuristic algorithms for prediction of daily diffuse solar radiation in air-polluted regions. *Renew. Energy* **2020**, *145*, 2034–2045. [CrossRef]

Article

Prediction of Building's Thermal Performance Using LSTM and MLP Neural Networks

Miguel Martínez Comesaña [1,*], Lara Febrero-Garrido [2], Francisco Troncoso-Pastoriza [1] and Javier Martínez-Torres [3]

1 Department of Mechanical Engineering, Heat Engines and Fluids Mechanics, Industrial Engineering School, University of Vigo, Maxwell s/n, 36310 Vigo, Spain; ftroncoso@uvigo.es
2 Defense University Center, Spanish Naval Academy, Plaza de España, s/n, 36920 Marín, Spain; lfebrero@cud.uvigo.es
3 Department of Applied Mathematics I. Telecommunications Engineering School, University of Vigo, 36310 Vigo, Spain; javmartinez@uvigo.es
* Correspondence: migmartinez@uvigo.es

Received: 22 September 2020; Accepted: 20 October 2020; Published: 23 October 2020

Abstract: Accurate prediction of building indoor temperatures and thermal demand is of great help to control and optimize the energy performance of a building. However, building thermal inertia and lag lead to complex nonlinear systems is difficult to model. In this context, the application of artificial neural networks (ANNs) in buildings has grown considerably in recent years. The aim of this work is to study the thermal inertia of a building by developing an innovative methodology using multi-layered perceptron (MLP) and long short-term memory (LSTM) neural networks. This approach was applied to a public library building located in the north of Spain. A comparison between the prediction errors according to the number of time lags introduced in the models has been carried out. Moreover, the accuracy of the models was measured using the CV(RMSE) as advised by AHSRAE. The main novelty of this work lies in the analysis of the building inertia, through machine learning algorithms, observing the information provided by the input of time lags in the models. The results of the study prove that the best models are those that consider the thermal lag. Errors below 15% for thermal demand and below 2% for indoor temperatures were achieved with the proposed methodology.

Keywords: neural network; LSTM; MLP; thermal inertia; building performance

---

1. Introduction

The residential sector makes an important contribution to energy consumption worldwide, representing more than the 40% of the total energy use in the European Union (EU) [1]. The EU has established several guidelines and directives to improve energy performance in buildings, such as 2010/31/EU (EPBD) [2] and 2012/27/EU [3], requiring that new buildings comply with nearly zero-energy buildings (NZEB) by 2030 [4] and to reach a decarbonized and highly energy efficient building stock by 2050. Therefore, energy efficiency in buildings is of great importance to the overall sustainability. Knowing exactly and precisely the energy consumption of a building is the first step to be able to optimize its energy performance. However, forecasting the building energy consumption is a difficult issue that many authors have investigated in recent years [5–7].

Different methods have been developed to predict building energy demand. Traditionally, dynamic simulation has been used successfully [8–10] becoming a suitable tool, that enables the assessment of building performance and the calculation of energy use. There are several building energy performance simulation (BEPS) tools available, such as TRNSYS [11], EnergyPlus [12] or DOE-2 [13]. However, the use of dynamic simulation tools requires the knowledge and control of many different parameters of the building, for example, envelope, materials properties, lighting, equipment, heating ventilation and air-conditioning (HVAC) systems and the behavior of the users, which makes the work of data collection arduous and increases the difficulty of obtaining exact results. The validation of these building energy simulation models is usually carried out with standard criteria such as the coefficient of variation of root mean square error (CV(RMSE)). Currently, these models are considered calibrated if they meet the criteria established by American Society of Heating Refrigerating and Air-Conditioning Engineers (ASHRAE) Guideline 14 [14], which states that a model is calibrated if its CV(RMSE) is below 15% monthly or 30% hourly. On the other hand, new tendencies in building prediction have emerged. Examples of this are the black box models, which stand out for their ability to train based on available data [15]. In this context, several data-driven models for building energy forecasting have been used more and more in recent years due to their robustness, resilience, strong yield and ease of implementation. Amongst them, the most popular data-driven approaches are the ones based on artificial neural networks (ANNs) [16,17].

ANNs are specific mathematical models that attempt to replicate the way a human neural network proceeds. The main and most important feature of these models is the ability to learn; that is, from known data, they are able to extract a pattern and then extrapolate the results to new data. Thus, neural networks have a remarkable ability to model the non-linear relationships between inputs and outputs, and for their massive interconnectivity [10,18,19]. Two of the most used ANNs in the context of building energy prediction are multi-layered perceptron (MLP) [19,20] and longshort-term memory (LSTM) neural networks [5,21]. MLP models are well known for being composed of several layers connecting the inputs to the specific output [10]. They have been applied to numerous scientific fields such as environment [22–24], econometrics [19,25], medical research [26], chemistry [27,28] and even building energy [10,29,30]. On the other hand, LSTMs are more current models, but their application has also expanded to a great number of fields. This type of ANN is characterized by a recurrent neural network (RNN) architecture [21,31]. RNNs differ from traditional feed-forward neural networks in that they have a hidden layer to consider connections with the previous values. This enables this type of neural network to model long-term dependencies; such as thermal inertia in buildings [32–34]. They encompass cyclic connections that make them, in principle, more suitable for modeling time sequence data than forward neural networks [35,36]. Some of the numerous fields where LSTM neural networks were applied are language modeling [36,37], speech recognition [35], tourism flow [38,39], sequential diagnosis [32,40] and energy analysis in buildings [21,41,42].

The aim of this work is to study the thermal inertia of a building by developing a methodology using MLP and LSTM neural networks. This procedure was applied to a public library building located in the north of Spain [43,44]. The available data are hourly observations of thermal demand and indoor temperatures of the building. Moreover, hourly observations of two weather variables (outdoor temperatures and solar radiation) and three time variables (hour of the day, day of the week and hour of the year) are also needed. A comparison between the prediction errors according to the number of time lags (each lag is one hour) introduced into the models, with respect to the studied variables, has been carried out. In addition, the accuracy of the models was measured using the CV(RMSE) as advised by AHSRAE. The novelty of this work lies in the analysis of the building inertia through two different artificial intelligence algorithms observing the information provided by the introduction of more time lags in the models.

2. Materials and Methods

2.1. Artificial Neural Networks Developed

The different mathematical models that have been built to carry out the analysis are presented in this section. The models are of two different types of ANNs: An MLP and an LSTM neural network.

2.1.1. Multi-Layered Perceptron Neural Networks

In this work the structure of the MLP neural network is variable due to the use of different numbers of time lags. Depending on the structure of these artificial neural networks, different network models can be generated. The most commonly used is the so-called feed-forward model. This model is composed of several layers (see Figure 1). The first layer, or input layer, is where the model inputs are introduced, whereas the last layer, or output layer, is where the results of the trained network are given. In addition, among them, the number of intermediate layers (hidden layers) can be zero, one or more [18,20]. The particular network architectures are differentiated by the number of hidden layers and hidden neurons according to the complexity of the problem [45,46]. The grid of tested values for the hidden layers was between 0 and 4. In contrast, the grid of values for the hidden neurons was composed of specific numbers based on the formulas presented by Shin-ike [47], Doukim et al. [48] or Vujicic et al. [49], which take into account the size of the sample and the number of inputs and outputs of the network. Furthermore, in this case, due to the complexity of the problem, this grid of neurons had to be expanded with the values: 50, 100, 200, 500 and 1000.

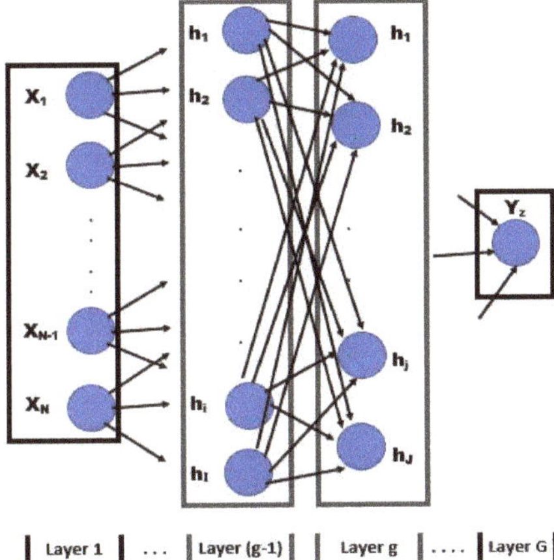

Figure 1. Multi-layered perceptron (MLP) neural network architecture with *N* inputs, *g* hidden layers and univariate output.

The MLP neural network training is done with a backward propagation algorithm; errors are propagated through the network and adapted to the hidden layers. Thus, error-correction learning (actual system responses must be known) is used to train the ANN [29,50]. There are several ways to perform the training [19,51] but, the methodology used in this work consists of updating the weights with an average update of the weights (batch learning), which is achieved by incorporating all the patterns in the input file (an epoch) and accumulating all the weight updates. There is also a need for a stop criterion [52,53]. Although there are other options, such as a threshold for the mean square error (MSE) or a limitation on the maximum number of iterations, the most widely used is cross-validation. This method is most effective in stopping training when the best generalization is achieved. It consists of separating a small part of the training data and using it to evaluate the trained network. Thus, training should stop the moment the network performance, quantified by the mean square error (MSE), begins to decrease or stagnates [18–20,31].

Lastly, in all MLP models created in this paper the activation function selected was the reLU function (rectified linear unit: $\max(0, x)$) [54], the kernel initializer was normal and the optimizing algorithm was the adaptive moment estimation (Adam) [55]. On the other hand, the batch size used to train the MLP neural networks was equal to 64 (mini-batch gradient descent [56]) and the patience (limit to stop the training if the performance of the model does not improve) was 100 epochs.

2.1.2. Long Short-Term Memory Neural Networks

LSTM is a recurrent neural network architecture specifically for modeling time sequences and their long-range dependencies more accurately than conventional RNNs. LSTM introduces specific units (memory blocks) into the recurrent hidden layer that store the temporal state of the network through self-connections. In addition, the memory block also contains units known as gates to control the flow of information [38,40]. A forget gate was incorporated into the memory blocks to prevent the LSTM model from processing continuous input flows without being segmented in subsequences. Furthermore, more complex LSTM networks also incorporate peephole connections between internal cells and their gates to learn the timing of the outputs [32,35].

The structure of traditional RNNs [57] can be presented by deterministic transitions from previous to current hidden states in form of a function:

$$h_t^{l-1},\ h_{t-1}^l \rightarrow h_t^l$$

where h_t^l represents a hidden state in layer l in timestep t.

However, LSTM counts with a complex dynamic that allows memorizing information for many timesteps. Long-term information is stored in a vector of memory cells $m_t^l \in \mathbb{R}^n$ [21,37]. The LSTM networks are empowered to decide whether to overwrite, retrieve or maintain this information for the next timestep. Their architecture is the following:

$$h_t^{l-1},\ h_{t-1}^l,\ m_{t-1}^l \rightarrow h_t^l,\ m_t^l$$

Let $x_1, x_2, \ldots, x_k$ be a classic input sequence (with k lags), where $x_t \in \mathbb{R}^D$ denotes the vector of D real values at the timestep t. The LSTM architecture defines an input gate i_t, a forget gate f_t and an output gate o_t. At specific time t, as is observable in Figure 2a where an LSTM cell is shown, the hidden layer output would be h_t, the cell input state $\widetilde{N}_t$ and the cell output state N_t [21,38,39,41].

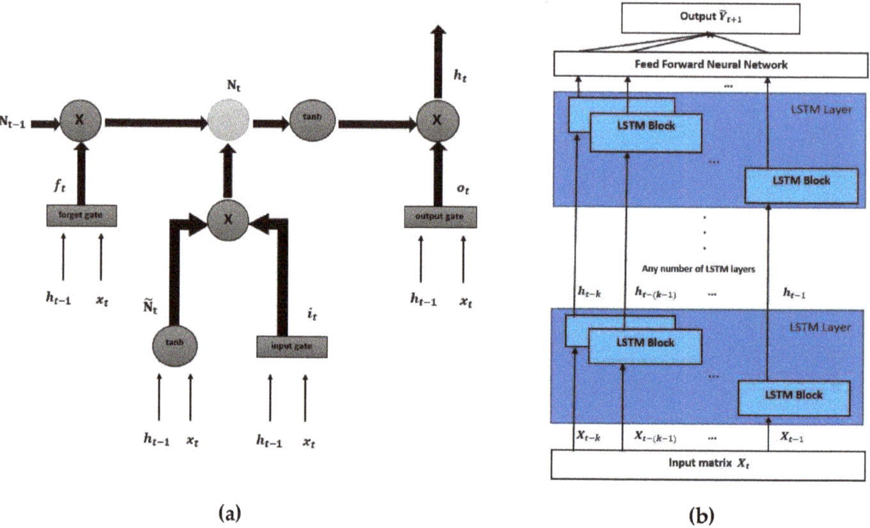

Figure 2. (**a**) Long short-term memory (LSTM) cell structure and (**b**) forecasting framework of LSTM neural network architecture.

The two values N_t (Equation (1)) and h_t (Equation (2)) that are transmitted to the next timestep, following the steps explained in [21,39], can be calculated as follows:

$$N_t = i_t \times \tilde{N}_t + f_t \times N_{t-1}, \quad (1)$$

$$h_t = o_t \times tanh(N_t) \quad (2)$$

A particular LSTM neural network was built for the prediction of indoor temperatures and thermal demand of the studied building (see Figure 2b). In this case, the input matrix of the network would be the observed data X_t (see Table 1) and the one-dimensional output the predicted future data $\tilde{Y}_{t+1}$ (Equation (3)). Once known h_t (Equation (2)) the network output will be:

$$\tilde{Y}_{t+1} = W_2 \times h_t + b, \quad (3)$$

where W_2 is the weight matrix connecting the last hidden layer to the output layer and b the bias term in the output layer (see Figure 2b).

Table 1. Summary of the data set, with the available variables, used in this study.

Date	Hour	Weekday	Yearhour	Outdoor Temp. (°C)	Radiation (W/m²)	Indoor Temp. (°C)	Thermal Demand (kWh)
05/02/2017 1:00	1	6	865	7.98	0.00	21.02	34.66
05/02/2017 2:00	2	6	866	8.28	0.00	20.99	36.85
05/02/2017 3:00	3	6	867	7.08	0.00	21.00	37.15
05/02/2017 4:00	4	6	868	5.98	0.00	21.04	38.78
05/02/2017 5:00	5	6	869	6.40	0.00	21.05	46.58
05/02/2017 6:00	6	6	870	6.42	0.00	21.03	37.81
05/02/2017 7:00	7	6	871	5.98	0.00	21.05	35.17
05/02/2017 8:00	8	6	872	6.52	1.17	21.06	33.07
05/02/2017 9:00	9	6	873	7.40	31.17	21.04	39.01
05/02/2017 10:00	10	6	874	7.70	180.17	21.04	42.19

Specifically, the LSTM architecture used in this study is also variable due to the different complexity of the models based on the number of time lags introduced. The variable parameters were the number of hidden and LSTM layers and the number of hidden neurons. In this case, the layer structures (LSTM and hidden layers) in the tested grid were: 2-1, 2-2, 3-1 and 3-2. On the other hand, the number of hidden neurons was a grid with eight values between 5 and 500. In addition, as in the MLP models, the activation function of the models was the reLU function [54], the kernel initializer was normal, the optimizing algorithm was Adam [55] and the batch size equal to 64 [56]. In order to avoid overfitting problems, they were trained with the same early stop as MLP neural networks (100 epochs) [52,53].

2.1.3. Pre-Processing Data

The data available in this study are hourly observations of two variables related to climate (outdoor temperature and solar radiation) and two related to the energy performance of the building (indoor temperature and thermal demand) between February and March of 2017 (see Table 1).

A continuous sample without large sets of missing values was necessary due to the use of hourly time lags as inputs in the models. This means that to train a model, besides the information of the independent variables at a given moment, the information of past instants of certain variables is also introduced. This study focuses on the prediction of thermal demand and indoor temperature of the studied building. The variables that have been used as explanatory variables of the models are presented in Figure 3. Although only the variables related to thermal conditions have been lagged, to add more information to the models, three time variables have been added (see Table 1). These three variables (hour of day, day of week and hour of year) do not provide any additional information being delayed due to their artificial nature.

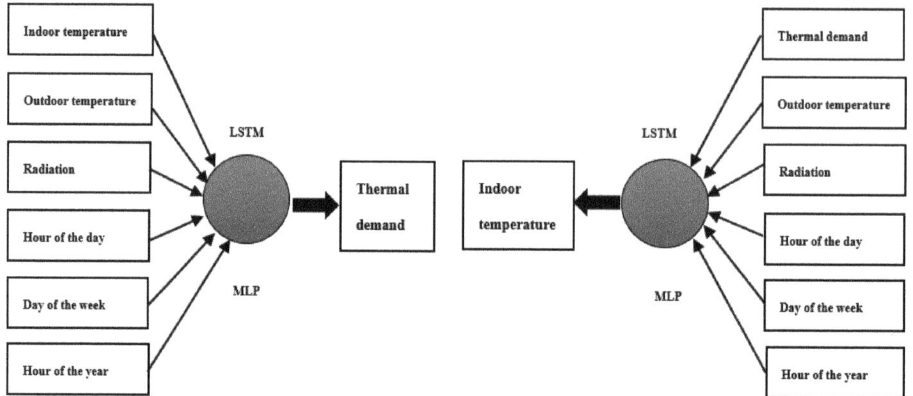

Figure 3. The variables used as input into the models to predict the thermal demand of the building (left) and the indoor temperatures of the building (right).

Generally, as the number of time lags considered increases, the complexity of the model needs to change (there is more information to be extracted). Therefore, a process based on repetition (cross validation design; data are divided into training, test and validation [58]) was carried out to find the best parameters for each model and for each of the different lagged architectures studied. This process consists of repeating a prediction of the same test sample 10 times with each model (the second week of January 2019), training with the same data, to extract the average performance of each of them and being able to compare them. Thus, the best architecture for each structure of lags and to each of the mathematical models used (LSTM or MLP) is obtained.

After finding the best model for each lag architecture analyzed (see Tables 2 and 3), the thermal inertia of the building has been tested with two new weeks of 2019 (validation data):

- Sample 1: 28/01/2019–03/02/2019
- Sample 2: 11/03/2019–17/03/2019

2.2. Data Acquisition of the Experimental Case Study

The process of data acquisition, as well as the characteristics of the energy system of the building studied, is described in this section.

2.2.1. Building and HVAC System Description

The case study building where this new methodology has been proved is a public library located in the city of Vigo, in the north west of Spain. It is a three floor building with connected and large open areas that trigger temperature stratification inside the library. The building has a working floor area of 820 m^2 and the average capacity is 220 persons. This total usable floor area includes 412.3 m^2 on the ground floor, 233.5 m^2 on the first floor and 73.0 m^2 on the second floor. Due to the interconnected spaces and the large window area, the building experiences high thermal inertia.

The HVAC system is a ground-source heat pump (GSHP) distributed with a radiant floor (RF). The reversible heat pump is a CIATESA IZA 185 with a nominal heating capacity of 45 kW and nominal cooling capacity of 35.7 kW, with a COP of 3.63 for heating and EER of 3.25 for cooling. It has a glycol-water mixture flow rate of 8450 m^3/h. The water storage tank contains 200 L. The GSHP is composed of six boreholes 100 m deep. The set point temperature for heating is 20 °C when the library is open and 17 °C when the library is closed. For cooling, the setpoint temperature is 26 °C when the library is open and 29 °C when closed. The tested heating period goes from October to May.

Further information about the building and its HVAC system can be consulted in [43,44,59].

2.2.2. Data Acquisition System

The data acquisition (DAQ) system in the building is further described in other articles by Cacabelos et al. [43,59] and Fernández et al. [44]. The rate of this system is one minute.

There are several wall module temperature sensors; six of them placed in the ground floor, four of them in the first floor and one in the second floor, all of them from the manufacturer Honeywell. In addition, the thermal zone temperature is calculated as the average of temperatures of every sensor from each floor. On the other hand, thermal energy consumption is measured with a thermal energy meter Multical 601 manufactured by Kamstrup.

The meteorological data were gathered from a weather station located at 42.17° N latitude and 8.68° E longitude at an altitude of 460 m. This station belongs to the weather station network from the Environment, Land Planning and Infrastructures Department, and it is located only 500 m from the building.

Further information about the DAQ and the weather station can be found in [43,44,59].

2.3. *Validation and Error Measurement*

The error measure considered in this paper to quantify the accuracy of each of the proposed models is the CV(RMSE) (coefficient of variation of the root mean square error):

$$CV(RMSE) = \frac{1}{\bar{Y}} \sqrt{\frac{\sum_{i=1}^{N}\left(Y_i - \hat{Y}_i\right)^2}{N}} \tag{4}$$

This measure was used both to find the best structure depending on the number of time lags introduced in the models and, also, to compare the different models with the test samples (Tables 2 and 3). It has been used in similar studies such as that of Hong et al. [60] and Kuo et al. [61].

3. Results and Discussion

The inertia in the thermal conditions of the Science Library of the University of Vigo was analyzed through the variation of the CV(RMSE) (Equation (4)) in the predictions of thermal demand and indoor temperatures of the building. In this analysis the models were trained with two months of hourly observations considering different numbers of time lags (values of certain variables in past hours) with respect to the variables of study to make the predictions. In this way, two specific weeks in 2019 (sample 1 and sample 2) have been taken into account to analyze the thermal inertia of the building by comparing the error obtained by each model according to the number of time lags used.

Section 3.1 presents the results to the thermal demand predictions and Section 3.2 the same analysis for indoor temperature predictions. In each section the evolution of the prediction error was analyzed based on the number of hourly time lags introduced in the models. The numerical results are the mean and the standard deviation of the CV(RMSE) obtained in 10 repetitions for each model and sample analyzed. All the specific architectures used in the models are shown in Tables 2 and 3. Lastly, all figures presented in this section were made with the Python programming language [62].

3.1. *Thermal Demand Analysis*

The results of the thermal demand predictions for the two samples studied, through which the thermal inertia of the building was analyzed in this section, are shown in Figure 4. Specifically, the results of the LSTM models are shown in Figure 4a and the results of the MLP models are shown in Figure 4b. Each time-lag structure analyzed is represented in the figures through the prediction with the lowest error among the 10 repetitions of the analysis.

LSTM models were capable of replicating the trend of the thermal demand of the building (except for certain peaks). They also yielded a small variation between the different replications of the analysis. This is shown in Table 2, with average errors of less than 20% and standard deviations below 0.05, in the two samples considered in this study. In sample 1 (first row of Figure 4a) the lag structure that produced the best results is the one with 6 lags (CV(RMSE) = 15.26%). In addition, it is also the structure with the smallest variation in the 10 experiment repetitions. As shown in Table 2 and Figure 4a, in this sample the average errors are, in general, higher than those of sample 2. Although in Figure 4a all the models are close to the real values, as only the best predictions of each model are shown, in Table 2 the differences between their average performances are presented. Thus, the model that considers 12 lags is the architecture that obtained the highest average error (CV(RMSE) = 19.13%) and presents the largest variability in its results. In sample 2 (second row of Figure 4a), where the models were better adjusted to reality, the architecture that obtained the best average performance, as shown in Table 2, is the 1-lag architecture (CV(RMSE) = 14.35%). In this case, smaller errors are not accompanied by more stable results; the highest stability was achieved with the model with 24 lags. On the contrary, the model that obtained the highest error and the largest variability is the one that considers 12 lags (CV(RMSE) = 17.34%).

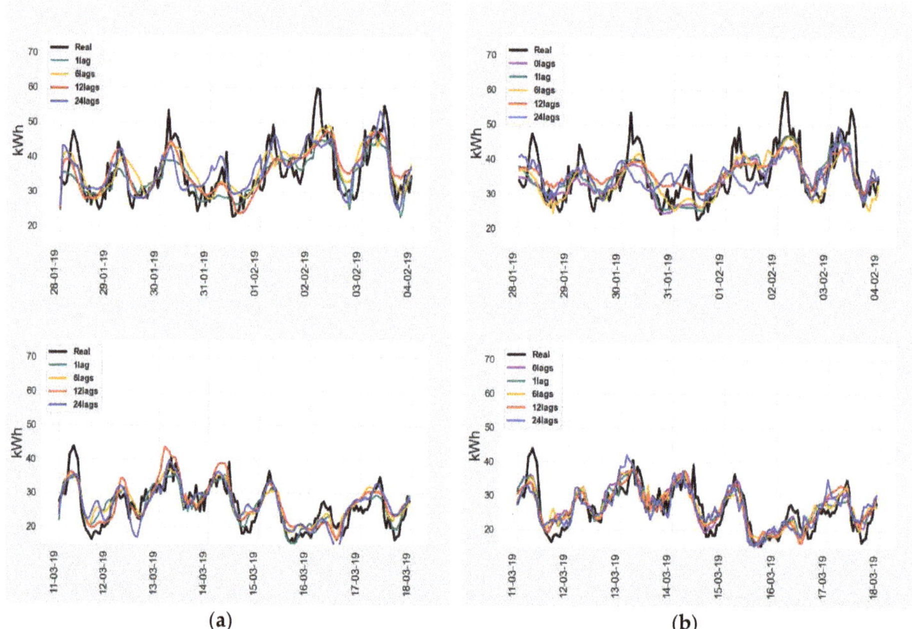

Figure 4. Results of thermal demand predictions in sample 1 (upper side) and sample 2 (bottom side) considering the different structures of time lags. Each architecture is represented by the prediction curve with the lowest error among the 10 repetitions: (**a**) LSTM models; (**b**) MLP models.

MLP models, as shown in Figure 4b, were also capable of predicting the actual thermal demand of the building with errors of less than 18%. As with LSTM models, certain peaks in the validation sample were not perfectly predicted and their variability among analysis repetitions was also small (below 0.02). However, these models yield better results than LSTM models (see Table 2). In sample 1 (first row of Figure 4b) the model with the lowest average error, as with LSTM models, is the model that considered 6 time lags (CV(RMSE) = 14.90%). In addition, this architecture is the one that produced the most stable results (see Table 2). On the contrary, the lag structure that adjusts reality in the worst way and produced the highest average error is the 12-lag structure (CV(RMSE) = 19.38%). It is also the structure that yields the most variable errors. In this case, a higher average error means more variability in predictions. In sample 2 (second row of Figure 4b), as in LSTM models, the predictions, on average, were better than in sample 1 (see Table 2). The architecture of lags that produced the lowest average error is the architecture of 1 lag (CV(RMSE) = 14.17%); the same structure as with the LSTM models. In this situation it is not the most stable model. On the other hand, the model that yielded the worst performance is the model that considered 24 lags (CV(RMSE) = 17.43%); not being the most variable model. In the case of sample 2, as presented in Table 2, the models with the least errors and the most stable models do not coincide.

Table 2. Results of the thermal demand predictions based on the time lags introduced in the neural network model. LSTM errors are presented in the first three columns, whereas MLP errors are presented in the last three. The results shown summarize the 10 repetitions of the experiment, for each architecture and each sample analyzed. The mean of the CV(RMSE) obtained in each of them and the standard deviation (SD) are presented. The different structures of the neural networks that obtain the best accuracy are presented between brackets.

LSTM Model	Sample 1		Sample 2		MLP Model	Sample 1		Sample 2	
	CV RMSE	SD	CV RMSE	SD		CV RMSE	SD	CV RMSE	SD
No lags	-	-	-	-	No lags (50-1)	16.05%	0.007	14.77%	0.005
1 lag (20-20-20-10-10-1)	18.07%	0.047	14.35%	0.035	1 lag (50-1)	16.27%	0.007	14.17%	0.008
6 lags (10-10-5-1)	15.26%	0.006	15.28%	0.012	6 lags (100-1)	14.90%	0.004	15.41%	0.010
12lags (10-10-10-5-5-1)	19.13%	0.066	17.34%	0.038	12lags (8-1)	16.27%	0.006	14.87%	0.004
24 lags (500-500-250-1)	15.90%	0.009	16.15%	0.009	24 lags (200-200-200-200-200-1)	19.38%	0.017	17.43%	0.007

In the case of the thermal demand of the building studied, its thermal inertia is significant but up to a certain limit of time lags. Both LSTM and MLP models show that the best fit to reality is obtained from an architecture with 6 lags in sample 1 and 1 lag in sample 2 (see Table 2). Therefore, the inertia influence is greater with the data of sample 1. Moreover, the LSTM models, for both samples, present the worst performance with 12 lags; but the result, although not surpassing the best one, improved by introducing 24 lags. On the other hand, with MLP models the introduction of more time lags than the optimal ones cause an increase in the average error. MLP models obtained the worst time-lag structure in both cases with the 24-lag structure. Thus, although the best MLP models are more accurate than the best LSTSM models, the worst MLP models are less accurate than the worst LSTM models. Regarding the stability of the results obtained, all the models show a small variability and, although there is not much difference between them, MLP models show smaller standard deviations. In addition, the model without time lags in any case obtained the best performance. This means that a thermal inertia exists and that the introduction of time lags in the thermal input variables provides valuable information.

3.2. Temperatures Analysis

The analysis of the thermal inertia of the building, for the two samples studied, through the predictions of the indoor temperatures of the building studied, is shown in Figure 5. Specifically, the results of the LSTM models are shown in Figure 5a and the results of the MLP models are shown in Figure 5b. As in the previous section, each time-lag structure studied is represented in the figures through the prediction with the lowest error among the 10 repetitions of the analysis.

LSTM models optimally predict the indoor temperatures of the building analyzed in this study. Table 3 shows that average errors (measured by the CV(RMSE)) are below 3% and their variability is small (below 0.005); both for sample 1 and for sample 2. In sample 1 (first row of Figure 5a), contrary to the previous section, the average errors are lower than those of sample 2. Furthermore, the lowest average error of this sample is obtained with the 1-lag structure (CV(RMSE) = 1.89%). In addition, although the models results are stable, this specific architecture shows the smallest standard deviation among all the structures analyzed (see Table 3). In this case, in Figure 5a it is observed that the two architectures with 12 and 24 lags are straight lines. This fact means that the introduction of more time lags does not provide useful information. Although the average error obtained by these two architectures is small, the models do not efficiently adjust to reality. Thus, the architecture that obtains the highest average error is the architecture with 24 lags (CV(RMSE) = 2.68%). Moreover, it is the least stable model structure in relation to the variability of the errors obtained (see Table 3). There is a correlation between the variability of the errors obtained and the average error; the greater the variability, the greater the average error. On the other hand, in sample 2 (second row of Figure 5a) the 1-lag structure, as in sample 1, obtained the lowest average error (CV(RMSE) = 2.05%) and the smallest variability (see Table 3). As shown in Figure 5(a), with this sample the different models analyzed show a more similar behavior among themselves than in sample 1. Furthermore, unlike the first sample, the introduction of more time lags in the models eventually reduces the average errors; the best models, are those with 1 and 24 time lags (see Table 3). Therefore, the architecture with the highest average error is the 6-lag architecture (CV(RMSE) = 2.23%). In this case, there is no relation between average error and the variability of the results.

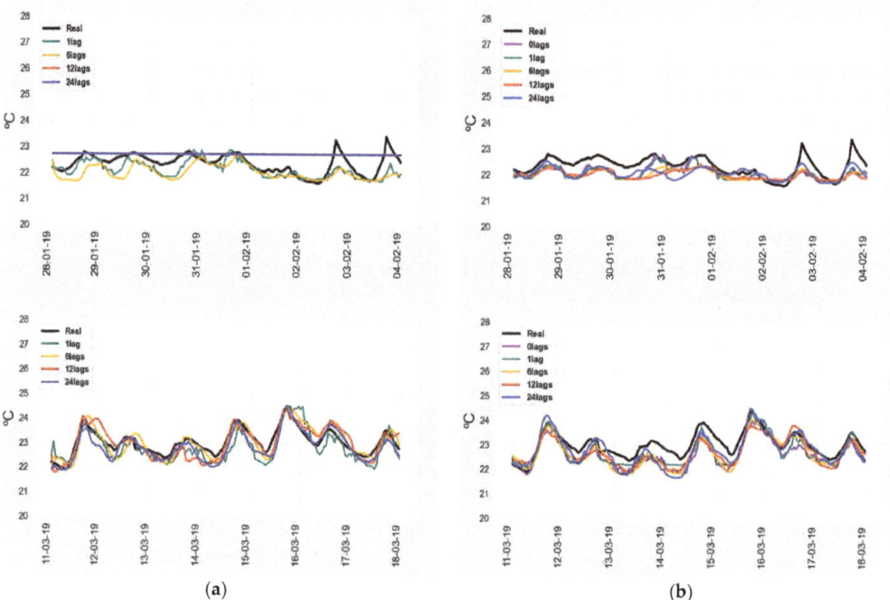

Figure 5. Results of the temperatures predictions in sample 1 (upper side) and sample 2 (bottom side) considering the different structures of time lags. Each architecture is represented by the prediction curve with the lowest error among the 10 repetitions: (**a**) LSTM models; (**b**) MLP models.

MLP models, presented in Figure 5b, once again provide better results than LSTM models. As shown in Table 3, their average errors are less than 2.5% and their variability is below 0.003. In sample 1 (first row of Figure 5b), the model structure with the lowest average error is the 24-lag structure (CV(RMSE) = 1.73%). On the contrary, the structure that produces the highest average error is the 6-lag structure (CV(RMSE) = 2.06%). In this case, and similarly to LSTM models, there is no direct relation between average error and variability in the results. In addition, as can be seen in the first row of Figure 5b (and the first row of Figure 5a), none of the models are able to replicate the peaks of the indoor temperatures at the weekend. Nevertheless, as shown in Table 3, the models do adjust the data trend efficiently. On the other hand, in sample 2 (second row of Figure 5b), as in LSTM models, the average errors obtained are higher than those of sample 1 (see Table 3). In this case, the architecture that yielded the lowest average error is the 12-lags architecture (CV(RMSE) = 1.90%). Additionally, it is also the architecture that presents the smallest variability among its results. With this sample, the introduction of time lags did provide valuable information because, as shown in Table 3, the best model structures are those with 12 and 24 lags. However, the lag structure that presents the highest average error was 6-lag structure (CV(RMSE) = 2.28%), slightly worse than the average error of the model without any lag. This means that, although the introduction of time lags provides useful information, the basic model (regardless of thermal inertia) already provides good results.

Table 3. Results of the indoor temperatures predictions based on the time lags introduced in the neural network model. LSTM errors are presented in the first three columns, whereas MLP errors are presented in the last three. The results shown summarize the 10 repetitions of the experiment, for each architecture and each sample analyzed. The mean of the CV(RMSE) obtained in each of them and the standard deviation (SD) are presented. The different structures of the neural networks that provide the best accuracy are presented between brackets.

LSTM Model	Sample 1		Sample 2		MLP Model	Sample 1		Sample 2	
	CV(RMSE)	SD	CV(RMSE)	SD		CV(RMSE)	SD	CV(RMSE)	SD
No lags	-	-	-	-	No lags (5-5-5-1)	1.82%	0.001	2.21%	≈ 0
1 lag (10-10-10-5-5-1)	1.89%	0.002	2.05%	0.001	1 lag (5-5-5-5-1)	1.86%	0.001	2.03%	0.003
6 lags (5-5-5-2-2-1)	2.40%	0.003	2.23%	0.004	6 lags (4-4-4-4-1)	2.06%	0.001	2.28%	0.003
12lags (5-5-5-2-1)	2.42%	0.002	2.19%	0.005	12lags (6-6-6-6-1)	2.02%	≈ 0	1.90%	≈ 0
24 lags (5-5-5-2-1)	2.68%	0.004	2.10%	0.004	24 lags (12-12-12-1)	1.73%	≈ 0	2.01%	0.001

Regarding the results of the indoor temperatures of the building, it can be observed that, on the one hand, there are less differences between the performance of the models in the different samples than in the case of the thermal demand. Moreover, in comparison with the thermal demand results, the different errors obtained for each of the architectures analyzed during the repetitions are lower and less variable. This demonstrates that indoor temperature values are more constant throughout a year (always around the set point temperature of the building). On the other hand, there are differences between LSTM and MLP models (see Table 3). With LSTM models the thermal inertia is less significant than in the case of thermal demand. In the two samples analyzed the best model was the one with a single time lag. In contrast, with MLP models the thermal inertia is more significant than in the analysis in the previous section. In both samples, the best model structures were a 24-lag structure for sample 1 and a 12-lag structure for sample 2; they use information from a higher number of past hours. In this case, MLP models show better performance in both the best and the worst case. In addition, even with less variability in general concerning indoor temperature predictions, MLP models again produced the lowest standard deviations. Finally, as in the case of thermal demand predictions, models without time lags do not produce the lowest average error in any of the analyses presented. This indicates that the thermal inertia is also significant in the case of indoor temperatures predictions.

4. Conclusions

A new application of machine and deep learning algorithms and a new methodology to analyze the influence of the thermal inertia of a building are presented in this paper. The study was conducted by analyzing monitored data of thermal demand and indoor temperatures of the Science Library of the University of Vigo. The data were completed with weather variables (outdoor temperatures and solar radiation) and three temporal variables (hour of the day, day of the week and hour of the year). The aim of this analysis is to study the importance of the thermal inertia of the building in predicting thermal demand and indoor temperatures. The methodology used in this study is based, on the one hand, on a classic machine learning model (MLP neural network) contrasting the improvement in the prediction accuracy provided by the thermal inertia of the building. On the other hand, as a comparison, a deep learning model (LSTM neural network) was carried out through an analogous analysis.

The research contribution of this paper is the application of mathematical methods to evaluate the influence of the thermal inertia of the building in the prediction errors of the thermal demand and the indoor temperatures of the building itself. The principal limitation of this research is the analysis of the thermal inertia of the building based on the difference in the errors rates of indoor temperatures depending on the number of time lags. The stable behavior of the indoor temperatures by default, produce that, in general, the prediction error is very low. This makes analysis and detection of the additional information that can provide the introduction of time lags in the models more difficult. On the other hand, the presented methods contribute with advantages over the already existing research in building simulation and thermal inertia analysis of buildings. With traditional building simulations models, both to make predictions and to analyze the thermal inertia of the building, it is necessary to have a deep knowledge about the specific subject and control many different energy parameters. With the methods presented in this paper, known as black box models, it is possible to do all kinds of analyses without any specific knowledge, and also to do it faster than dynamic simulation methods. These models rely on having a significant amount of data to be able to extract information from it and extrapolate it. Additionally, the variables involved in this study are commonly monitored in numerous buildings. Thus, the methodology presented can be applied to evaluate the energy performance or the influence of thermal inertia of different buildings.

First, the results show that the black box models yield average prediction errors below the proposed values for considering calibrated models. The results also illustrate the greater accuracy of MLP neural networks in predicting both thermal demand and indoor temperatures. LSTM neural networks, created specifically to model time sequences, show a poorer performance than MLP models due to the peculiarity of this analysis. In this study the models were trained with only two months of data to try to predict future time sequences of the variable of interest. LSTM models are focused on analyses where predictions are made for time sequences immediately consecutive to the training data (as a time series). In this specific case, where the data are not continuous in time, LSTM models are not the most efficient. In this paper it was demonstrated that, for analyses similar to the one presented here, MLP models are more appropriate; they are more flexible if the available data, on the one hand, are not continuous in time or, on the other hand, are very few. In terms of thermal demand predictions, for both sample 1 and sample 2, the best MLP models are better than the best LSTM models. While the average errors of the MLP models were 14.90% and 14.17%, respectively, the average errors of the LSTM models were 15.26% and 14.35%. The same applies to indoor temperature predictions. The average errors of the best MLP models were 1.73% in sample 1 and 1.90% in sample 2, and those of the best LSTM models were 1.89% and 2.05%, respectively. Regarding the variability of the predictions, the results of the LSTM and MLP models are similar, but the MLP models, in most cases, have slightly smaller standard deviations.

Furthermore, the results of the thermal inertia analysis of the building are different in relation to the different models used. Although all the analyses show that the introduction of specific time lags reduces the prediction error, the influence of the thermal inertia of the building changes with LSTM and MLP models. In the case of thermal demand analysis, both LSTM and MLP models show the same result. In sample 1 the optimal number of time lags introduced in the model was 6 while in sample 2 the optimal structure was a model with 1 time lag. It was demonstrated that the inertia of the independent variables of these models (indoor temperatures, outdoor temperatures and solar radiation) does not extend too much in time. Differently, the results of LSTM and MLP models do not coincide with the indoor temperatures analysis. LSTM models, both of sample 1 and sample 2, have their optimum in 1 time-lag. In contrast, with MLP models the best structures were 24 time lags in sample 1 and 12 time lags in sample 2. Thus, MLP models are able to better extract the information provided by the introduction of more time lags; they are able to exploit the thermal inertia of the building.

From an energy point of view, the conclusion is that the thermal inertia of the building does provide useful information to the machine learning models by introducing lagged variables. This was demonstrated by the decrease, up to a certain limit, of average errors with the introduction of time lags. Therefore, it was also demonstrated that the dimension of the thermal inertia, measured by the maximum number of time lags that provides valuable information, can be detected by the methods presented in this paper. Moreover, the analysis showed that MLP models are more appropriated than LSTM models if the available data set is not large and continuous over time. For the two variables and the two samples analyzed the average error produced by the optimal MLP models are lower than those provide by the optimal LSTM models. Lastly, it was shown that there is more inertia in thermal demand data than in indoor temperatures data; the introduction of thermal demand lags, due to the higher variability of their data, provides more useful information to the models than indoor temperature ones.

Author Contributions: Conceptualization, M.M.C. and L.F.-G.; methodology, M.M.C. and J.M.-T.; software, M.M.C. and F.T.-P.; validation, L.F.-G., J.M.-T. and F.T.-P.; formal analysis, M.M.C. and L.F.-G.; investigation, M.M.C. and L.F.-G.; resources, J.M.-T. and L.F.-G.; data curation, F.T.-P.; writing—original draft preparation, M.M.C. and L.F.-G.; writing—review and editing, J.M.-T. and F.T.-P.; visualization, M.M.C. and F.T.-P.; supervision, L.F.-G., J.M.-T. and J.T.-P.; project administration, M.M.C. and L.F.-G. All authors have read and agreed to the published version of the manuscript.

Funding: This research was funded by the Spanish Government (Science, Innovation and Universities Ministry) under the project RTI2018-096296-B-C21.

Acknowledgments: This research was supported by the Spanish Government (Science, Innovation and Universities Ministry) under the project RTI2018-096296-B-C21.

Conflicts of Interest: The authors declare no conflict of interest.

References

1. Li, D.H.W.; Yang, L.; Lam, J.C. Zero energy buildings and sustainable development implications—A review. *Energy* **2013**, *54*, 1–10. [CrossRef]
2. Official Journal of the European Union. *Directive 2010/31/EU of the European Parliament and of the Council of 19 May 2010 on the Energy Performance of Buildings*; EU: Brussels, Belgium, 2010.
3. European Commission. *Directive 2012/27/EU of the European Parliament and of the Council of 25 October 2012 on Energy Efficiency, Amending DIRECTIVES 2009/125/EC and 2010/30/EU and Repealing Directives 2004/8/EC and 2006/32/EC.*; EU: Brussels, Belgium, 2012.
4. Nematchoua, M.K.; Marie-Reine Nishimwe, A.; Reiter, S. Towards nearly zero-energy residential neighbourhoods in the European Union: A case study. *Renew. Sustain. Energy Rev.* **2021**, *135*, 110198. [CrossRef]
5. Bourdeau, M.; Zhai, X.q.; Nefzaoui, E.; Guo, X.; Chatellier, P. Modeling and forecasting building energy consumption: A review of data-driven techniques. *Sustain. Cities Soc.* **2019**, *48*, 101533. [CrossRef]
6. Deb, C.; Zhang, F.; Yang, J.; Lee, S.E.; Shah, K.W. A review on time series forecasting techniques for building energy consumption. *Renew. Sustain. Energy Rev.* **2017**, *74*, 902–924. [CrossRef]
7. Fumo, N.; Mago, P.; Luck, R. Methodology to estimate building energy consumption using EnergyPlus Benchmark Models. *Energy Build.* **2010**, *42*, 2331–2337. [CrossRef]
8. Harish, V.S.K.V.; Kumar, A. A review on modeling and simulation of building energy systems. *Renew. Sustain. Energy Rev.* **2016**, *56*, 1272–1292. [CrossRef]
9. Heiple, S.; Sailor, D.J. Using building energy simulation and geospatial modeling techniques to determine high resolution building sector energy consumption profiles. *Energy Build.* **2008**, *40*, 1426–1436. [CrossRef]
10. Neto, A.H.; Fiorelli, F.A.S. Comparison between detailed model simulation and artificial neural network for forecasting building energy consumption. *Energy Build.* **2008**, *40*, 2169–2176. [CrossRef]
11. University of Wisconsin—Madison. *Solar Energy, L. TRNSYS, a Transient Simulation Program*; The Laboratory: Madison, WI, USA, 1975.
12. Crawley, D.B.; Lawrie, L.K.; Winkelmann, F.C.; Buhl, W.F.; Huang, Y.J.; Pedersen, C.O.; Strand, R.K.; Liesen, R.J.; Fisher, D.E.; Witte, M.J.; et al. EnergyPlus: Creating a new-generation building energy simulation program. *Energy Build.* **2001**, *33*, 319–331. [CrossRef]
13. James, J. Hirsch & Associates (JJH). DOE-2 (version 2.2-047d). Available online: http://www.doe2.com/ (accessed on 1 September 2020).
14. ASHRAE. *Guideline 14-2014—Measurement of Energy, Demand, and Water Savings*; ASHRAE: Atlanta, GA, USA, 2014.
15. Guidotti, R.; Monreale, A.; Ruggieri, S.; Turini, F.; Giannotti, F.; Pedreschi, D. A Survey of Methods for Explaining Black Box Models. *ACM Comput. Surv.* **2018**, *51*, 93. [CrossRef]
16. Runge, J.; Zmeureanu, R. Forecasting Energy Use in Buildings Using Artificial Neural Networks: A Review. *Energies* **2019**, *12*, 3254. [CrossRef]
17. Oh, S. Comparison of a Response Surface Method and Artificial Neural Network in Predicting the Aerodynamic Performance of a Wind Turbine Airfoil and Its Optimization. *Appl. Sci.* **2020**, *10*, 6277. [CrossRef]

18. Hung, N.Q.; Babel, M.S.; Weesakul, S.; Tripathi, N.K. An artificial neural network model for rainfall forecasting in Bangkok, Thailand. *Hydrol. Earth Syst. Sci.* **2009**, *13*, 1413–1425. [CrossRef]
19. Guresen, E.; Kayakutlu, G.; Daim, T.U. Using artificial neural network models in stock market index prediction. *Expert Syst. Appl.* **2011**, *38*, 10389–10397. [CrossRef]
20. Fadare, D.A. Modelling of solar energy potential in Nigeria using an artificial neural network model. *Appl. Energy* **2009**, *86*, 1410–1422. [CrossRef]
21. Kong, W.; Dong, Z.Y.; Jia, Y.; Hill, D.J.; Xu, Y.; Zhang, Y. Short-Term Residential Load Forecasting Based on LSTM Recurrent Neural Network. *IEEE Trans. Smart Grid* **2019**, *10*, 841–851. [CrossRef]
22. Banerjee, P.; Singh, V.S.; Chatttopadhyay, K.; Chandra, P.C.; Singh, B. Artificial neural network model as a potential alternative for groundwater salinity forecasting. *J. Hydrol.* **2011**, *398*, 212–220. [CrossRef]
23. Iglesias, C.; Martínez Torres, J.; García Nieto, P.J.; Alonso Fernández, J.R.; Díaz Muñiz, C.; Piñeiro, J.I.; Taboada, J. Turbidity Prediction in a River Basin by Using Artificial Neural Networks: A Case Study in Northern Spain. *Water Resour. Manag.* **2014**, *28*, 319–331. [CrossRef]
24. Anjos, O.; Iglesias, C.; Peres, F.; Martínez, J.; García, Á.; Taboada, J. Neural networks applied to discriminate botanical origin of honeys. *Food Chem.* **2015**, *175*, 128–136. [CrossRef]
25. Gil-Cordero, E.; Cabrera-Sánchez, J.-P. Private Label and Macroeconomic Indexes: An Artificial Neural Networks Application. *Appl. Sci.* **2020**, *10*, 6043. [CrossRef]
26. Nasser, I.M.; Abu-Naser, S.S. Predicting Tumor Category Using Artificial Neural Networks. *Int. J. Acad. Health Med Res. (IJAHMR)* **2019**, *3*, 1–7.
27. Curteanu, S.; Cartwright, H. Neural networks applied in chemistry. I. Determination of the optimal topology of multilayer perceptron neural networks. *J. Chemom.* **2011**, *25*, 527–549. [CrossRef]
28. Iglesias, C.; Anjos, O.; Martínez, J.; Pereira, H.; Taboada, J. Prediction of tension properties of cork from its physical properties using neural networks. *Eur. J. Wood Wood Prod.* **2015**, *73*, 347–356. [CrossRef]
29. Chae, Y.T.; Horesh, R.; Hwang, Y.; Lee, Y.M. Artificial neural network model for forecasting sub-hourly electricity usage in commercial buildings. *Energy Build.* **2016**, *111*, 184–194. [CrossRef]
30. Robinson, C.; Dilkina, B.; Hubbs, J.; Zhang, W.; Guhathakurta, S.; Brown, M.A.; Pendyala, R.M. Machine learning approaches for estimating commercial building energy consumption. *Appl. Energy* **2017**, *208*, 889–904. [CrossRef]
31. Rahman, A.; Smith, A.D. Predicting heating demand and sizing a stratified thermal storage tank using deep learning algorithms. *Appl. Energy* **2018**, *228*, 108–121. [CrossRef]
32. Lipton, Z.; Kale, D.; Elkan, C.; Wetzel, R. Learning to Diagnose with LSTM Recurrent Neural Networks. *arXiv* **2015**, arXiv:1511.03677.
33. Mikolov, T.; Zweig, G. Context Dependent Recurrent Neural Network Language Model. In Proceedings of the 2012 IEEE Workshop on Spoken Language Technology, SLT 2012, Miami, FL, USA, 2–5 December 2012. [CrossRef]
34. Poulose, A.; Han, D.S. UWB Indoor Localization Using Deep Learning LSTM Networks. *Appl. Sci.* **2020**, *10*, 6290. [CrossRef]
35. Sak, H.; Senior, A.; Beaufays, F. Long Short-Term Memory Based Recurrent Neural Network Architectures for Large Vocabulary Speech Recognition. *arXiv* **2014**, arXiv:1402.1128.
36. Sundermeyer, M.; Ney, H.; Schluter, R. From Feedforward to Recurrent LSTM Neural Networks for Language Modeling. *Audio Speech Lang. Process. IEEE/ACM Trans.* **2015**, *23*, 517–529. [CrossRef]
37. Sundermeyer, M.; Schlüter, R.; Ney, H. *LSTM Neural Networks for Language Modeling*; Science Department RWTH Aachen University: Aachen, Germany, 2012.
38. Li, Y.; Cao, H. Prediction for Tourism Flow based on LSTM Neural Network. *Procedia Comput. Sci.* **2018**, *129*, 277–283. [CrossRef]
39. Duan, Y.; Yisheng, L. Travel Time Prediction with LSTM Neural Network. In Proceedings of the 2016 IEEE 19th International Conference on Intelligent Transportation Systems (ITSC), Rio De Janeiro, Brazil, 1–4 November 2016; pp. 1053–1058. [CrossRef]
40. Zhao, H.; Sun, S.; Jin, B. Sequential Fault Diagnosis Based on LSTM Neural Network. *IEEE Access* **2018**, *6*, 12929–12939. [CrossRef]
41. Kim, T.-Y.; Cho, S.-B. Predicting residential energy consumption using CNN-LSTM neural networks. *Energy* **2019**, *182*, 72–81. [CrossRef]

42. Marino, D.L.; Amarasinghe, K.; Manic, M. Building energy load forecasting using Deep Neural Networks. In Proceedings of the IECON 2016—42nd Annual Conference of the IEEE Industrial Electronics Society, Florence, Italy, 23–26 October 2016; pp. 7046–7051.
43. Cacabelos, A.; Eguía, P.; Míguez, J.L.; Granada, E.; Arce, M.E. Calibrated simulation of a public library HVAC system with a ground-source heat pump and a radiant floor using TRNSYS and GenOpt. *Energy Build.* **2015**, *108*, 114–126. [CrossRef]
44. Fernández, M.; Eguía, P.; Granada, E.; Febrero, L. Sensitivity analysis of a vertical geothermal heat exchanger dynamic simulation: Calibration and error determination. *Geothermics* **2017**, *70*, 249–259. [CrossRef]
45. Sheela, K.G.; Deepa, S.N. Review on Methods to Fix Number of Hidden Neurons in Neural Networks. *Math. Probl. Eng.* **2013**, *2013*, 425740. [CrossRef]
46. Chen, K.; Yang, S.; Batur, C. Effect of multi-hidden-layer structure on performance of BP neural network: Probe. In Proceedings of the 2012 8th International Conference on Natural Computation, Chongqing, China, 29–31 May 2012; pp. 1–5.
47. Shin-ike, K. A two phase method for determining the number of neurons in the hidden layer of a 3-layer neural network. In Proceedings of the SICE Annual Conference 2010, Taipei, Taiwan, 18–21 August 2010; pp. 238–242.
48. Doukim, C.; Dargham, J.; Chekima, A. Finding the Number of hidden Neurons for an MLP Neural Network Using Coarse to Fine Search Technique. In Proceedings of the 10th International Conference on Information Science, Signal Processing and their Applications (ISSPA 2010), Kuala Lumpur, Malaysia, 10–13 May 2010; pp. 606–609. [CrossRef]
49. Vujicic, T.; Matijević, T.; Ljucovic, J.; Balota, A.; Sevarac, Z. Comparative Analysis of Methods for Determining Number of Hidden Neurons in Artificial Neural Network. *Artif. Intell. Rev.* **2016**, *48*. [CrossRef]
50. Pradhan, B.; Lee, S. Landslide risk analysis using artificial neural network model focusing on different training sites. *Int. J. Phys. Sci.* **2009**, *4*, 1–15.
51. Nakama, T. Theoretical analysis of batch and on-line training for gradient descent learning in neural networks. *Neurocomputing* **2009**, *73*, 151–159. [CrossRef]
52. Li, M.; Soltanolkotabi, M.; Oymak, S. Gradient Descent with Early Stopping is Provably Robust to Label Noise for Overparameterized Neural Networks. In Proceedings of the Twenty Third International Conference on Artificial Intelligence and Statistics, Palermo, Italy, 3–5 June 2020; pp. 4313–4324.
53. Bilbao, I.; Bilbao, J. Overfitting problem and the over-training in the era of data: Particularly for Artificial Neural Networks. In Proceedings of the 2017 Eighth International Conference on Intelligent Computing and Information Systems (ICICIS), Cairo, Egypt, 5–7 December 2017; pp. 173–177.
54. Eckle, K.; Schmidt-Hieber, J. A comparison of deep networks with ReLU activation function and linear spline-type methods. *Neural Netw.* **2019**, *110*, 232–242. [CrossRef]
55. Bock, S.; Weiß, M. A Proof of Local Convergence for the Adam Optimizer. In Proceedings of the 2019 International Joint Conference on Neural Networks (IJCNN), Budapest, Hungary, 14–19 July 2019; pp. 1–8.
56. Li, M.; Zhang, T.; Chen, Y.; Smola, A. Efficient mini-batch training for stochastic optimization. In Proceedings of the ACM SIGKDD International Conference on Knowledge Discovery and Data Mining, New York, NY, USA, 24–27 August 2014. [CrossRef]
57. Zaremba, W.; Sutskever, I.; Vinyals, O. Recurrent Neural Network Regularization. *arXiv* **2014**, arXiv:1409.2329.
58. Singh, G.; Panda, R. Daily Sediment Yield Modeling with Artificial Neural Network using 10-fold Cross Validation Method: A small agricultural watershed, Kapgari, India. *Int J. Earth Sci Eng* **2011**, *4*, 443–450.
59. Cacabelos, A.; Eguía, P.; Febrero Garrido, L.; Granada, E. Development of a new multi-stage building energy model calibration methodology and validation in a public library. *Energy Build.* **2017**, *146*. [CrossRef]
60. Hong, T.; Kim, C.-J.; Jaemin, J.; Kim, J.; Koo, C.; Jeong, K.; Lee, M. Framework for Approaching the Minimum CV(RMSE) using Energy Simulation and Optimization Tool. *Energy Procedia* **2016**, *88*, 265–270. [CrossRef]
61. Kuo, P.-H.; Huang, C.-J. A High Precision Artificial Neural Networks Model for Short-Term Energy Load Forecasting. *Energies* **2018**, *11*, 213. [CrossRef]
62. Pilgrim, M.; Willison, S. *Dive into Python 3*; Springer: Berlin, Germany, 2009; Volume 2.

Publisher's Note: MDPI stays neutral with regard to jurisdictional claims in published maps and institutional affiliations.

Article

Heat Loss Coefficient Estimation Applied to Existing Buildings through Machine Learning Models

Miguel Martínez-Comesaña [1,*], Lara Febrero-Garrido [2], Enrique Granada-Álvarez [1], Javier Martínez-Torres [3] and Sandra Martínez-Mariño [1]

1. Department of Mechanical Engineering, Heat Engines and Fluids Mechanics, Industrial Engineering School, University of Vigo, Maxwell s/n, 36310 Vigo, Spain; egranada@uvigo.es (E.G.-Á.); samartinez@uvigo.es (S.M.-M.)
2. Defense University Center, Spanish Naval Academy, Plaza de España, s/n, 36920 Marín, Spain; lfebrero@cud.uvigo.es
3. Department of Applied Mathematics I, Telecommunications Engineering School, University of Vigo, 36310 Vigo, Spain; javmartinez@uvigo.es

* Correspondence: migmartinez@uvigo.es

Received: 23 November 2020; Accepted: 12 December 2020; Published: 16 December 2020

Abstract: The Heat Loss Coefficient (HLC) characterizes the envelope efficiency of a building under in-use conditions, and it represents one of the main causes of the performance gap between the building design and its real operation. Accurate estimations of the HLC contribute to optimizing the energy consumption of a building. In this context, the application of black-box models in building energy analysis has been consolidated in recent years. The aim of this paper is to estimate the HLC of an existing building through the prediction of building thermal demands using a methodology based on Machine Learning (ML) models. Specifically, three different ML methods are applied to a public library in the northwest of Spain and compared; eXtreme Gradient Boosting (XGBoost), Support Vector Regression (SVR) and Multi-Layer Perceptron (MLP) neural network. Furthermore, the accuracy of the results is measured, on the one hand, using both CV(RMSE) and Normalized Mean Biased Error (NMBE), as advised by AHSRAE, for thermal demand predictions and, on the other, an absolute error for HLC estimations. The main novelty of this paper lies in the estimation of the HLC of a building considering thermal demand predictions reducing the requirement for monitoring. The results show that the most accurate model is capable of estimating the HLC of the building with an absolute error between 4 and 6%.

Keywords: energy efficiency; heat loss coefficient; machine learning; XGBoost; MLP; SVR

---

1. Introduction

Energy efficiency in buildings is a key roadmap to achieve sustainable development worldwide. Nearly one third of the primary energy consumed in the world comes from non-industrial buildings, and it is similar to all the energy used in the transport sector [1]. Consequently, the residential sector constitutes the largest contributor to global warming by means of carbon dioxide emissions. To achieve the goal of reducing building energy consumption, international cooperation is crucial. The European Union (EU) has created a strict legislative framework implementing two important directives: Energy Performance of Buildings Directive 2010/31/EU (EPBD) [2] and the Energy Efficiency Directive 2012/27/EU [3]. Both of them were updated and amended over the years. Specifically, the Directive amending the Energy Performance of Buildings Directive 2018/844/EU [4] includes different aspects strengthening the commitment of real building stock modernisation through technological improvements [5]. On the other hand, the International Energy Agency (IEA) is also working hard on this issue by means of its Energy in Buildings and Community (EBC) research programme developing activities towards Nearly Zero Energy Buildings (NZEB), the reduction of carbon dioxide emissions and energy savings technologies. One of the high priority research project themes is the building envelope.

The envelope of a building is one of the main sources of heat transfer between the exterior and the interior of the rooms, and it is responsible for the difference between the conditions predicted at the design stage and the actual conditions of the use of the building, including lighting, occupancy and Heating, Ventilation and Air Conditioning (HVAC) systems' operation. This is called the performance gap between building design and operation [6,7]. The envelope efficiency of a building can be characterized under in-use conditions by calculating the Heat Loss Coefficient (HLC) [8]. This coefficient determines the rate of heat flow through the buildings' envelope when a temperature difference exists between the indoor air and the outdoor air under steady state conditions [9]. Therefore, the accurate estimation of the HLC contributes to optimizing the energy consumption of a building. On the other hand, building energy performance simulation tools have been used in recent years to analyse the energy behaviour of buildings [10]. However, achieving an accurate simulation is not trivial. Many sources of error are introduced into the simulation: occupancy and user's behaviour, weather data [11], envelope materials and thickness, the use of electric equipment, etc. The alternative to simulation is the monitoring of the building operation, which leads to the difficulty of the installation of sensors in in-use buildings. Therefore, in this context, the application of advanced mathematical modelling techniques, such as machine learning methods, is becoming more and more common [12].

There are three widely used building energy prediction models: the white-box (physics-based), black-box (data-driven) and grey-box (a combination of physics based and data-driven) modelling approaches [13]. The application of black-box models in building energy analysis has been consolidated in recent years. These models have received considerable attention because of their great success in learning complex patterns and being able to make accurate predictions without specific knowledge of the subject [14,15]. Machine Learning (ML) models, also known as black-box models, are specific mathematical models that are capable of learning a pattern from data and extrapolating it to a new sample. Thus, they were widely applied in studies presenting control strategies to reduce energy costs. Three of the most used are: eXtreme Gradient Boosting (XGBoost), Support Vector Regression (SVR) and Multi-Layer Perceptron (MLP) neural network. Among them, XGBoost is characterized for building regression trees, one-by-one, so that the subsequent models are trained taking into account the residuals of the previous ones [16]. This algorithm has been used in numerous fields such as biology [17,18], econometrics [19], the environment [20,21] or, as in this case, the energy performance of buildings [22,23]. SVR models emerged as a predictive alternative due to the use of a distinctive loss function [24,25], on the one hand, and the dual formulation of the problem [26], on the other. This algorithm focuses on minimising an upper bound of the generalization error instead of minimising the prediction error in the training sample (*empirical risk minimization*) [24]. The usefulness of this algorithm is demonstrated by its expansion into scientific fields such as econometrics [26], health [27,28], electrical efficiency in cities [29], as well as energy consumption in buildings [25,30]. Lastly, MLP neural networks have stood out in past years for their great capacity to model non-linear relationships between certain inputs and outputs, as well as for their massive interconnectivity [31–33]. Furthermore, this type of Artificial Neural Network (ANN) has been applied to different areas of study such as chemistry [34], the environment [35,36], sensors [37] or energy analysis of buildings [38–40].

The aim of this paper is to estimate the HLC of an existing building through the prediction of building thermal demands using a methodology based on ML models. Specifically, three different methods are applied to a public library in the northwest of Spain and compared: XGBoost, SVR and MLP neural network. The dataset consists of hourly observations of two climate variables (outdoor temperatures and solar radiation) and two variables related to the thermal behaviour of the building (heating demand and indoor temperatures). In addition, to improve model training, three temporal variables (hour of the year, day of the week and hour of the day) are taken into account as model inputs. A comparison between the errors obtained by the different machine learning models is carried out on three different validation samples. Moreover, the accuracy of each of the models is quantified with CV(RMSE) and Normalized Mean Biased Error (NMBE), both recommended by the American Society of Heating, Refrigerating and Air-Conditioning Engineers (ASHRAE), and their usefulness was demonstrated in [41,42]. The novelty of this work lies in the application of machine learning models to estimate the HLC of a building through heating demand predictions. In this context, this paper contributes with a methodology that reduces the exigency of building monitoring, and heating demands will be accurately predicted. Additionally, this methodology does not depend on physical simulations, which require huge amounts of input data and specific knowledge and which have uncertainty in the evaluation of energy renovation or the energy certification of buildings.

2. Materials and Methods

2.1. Heat Loss Coefficient Calculation

The HLC is the most used Key Performance Indicator (KPI) to describe the building envelope energy efficiency. This coefficient reflects transmission heat losses through the envelope per degree difference between indoor and outdoor temperatures from walls, roofs and floors (UA(kW/K)) and ventilation and/or infiltration heat losses per degree difference between indoor and outdoor temperatures (Cv(kW/K))) [43]. Following the method developed by Uriarte et al. [8], the HLC can be described as Equation (1):

$$HLC = (UA + C_v) \quad (\mathrm{kW/K}) \tag{1}$$

where UA stands for the building envelope transmission heat transfer coefficient (kW/K) and Cv stands for the infiltration and/or ventilation heat loss coefficient (kW/K). Moreover, the energy balance used to calculate the HLC, assuming stationary conditions, was developed by Uriate et al. [8,44], and it is presented in Equation (2):

$$\sum_{k=1}^{N} Q_k + \sum_{k=1}^{N} K_k = HLC \sum_{k=1}^{N} (T_{in,k} - T_{out,k}) - \sum_{k=1}^{N} (S_a V_{sol})_k \tag{2}$$

where Q_k represents the heating gains (kW), K_k represents the internal gains (kW) caused by occupants and electricity consumption ($K_k = K_{electricity,k} + K_{occupancy,k}$), $(T_{in,k} - T_{out,k})$ summarizes the gap between temperatures inside and outside the building in degrees Kelvin and $S_a V_{sol}$ stands for the solar gains (kW).

Several terms of this heat exchange are difficult to measure for in-use buildings. For example, the solar gains are complicated to know, and therefore, they introduce an important source of uncertainty into the equation. Thus, in this study, only cold and cloudy periods are considered where solar radiation was very low; direct solar radiation can be considered null; and all the radiation can be perceived as purely diffuse [45]. This is why in this study, direct solar radiation is not taken into account, and only diffuse solar radiation is considered. Another parameter that is difficult to estimate is the occupancy. Therefore, with the aim of reducing the sources of uncertainty, the study is developed during weekends when there is no occupancy. Then, the result of Equation (2) will give an idea of how well built the building is in terms of the enclosure of its envelope.

In this way, the HLC formula that is used in this analysis (average method), taking into account that the periods analysed in this paper were weekends with no occupation and in which there was only purely diffuse radiation, is shown in Equation (3):

$$HLC = \frac{\sum_{k=1}^{N} Q_k}{\sum_{k=1}^{N} (T_{in,k} - T_{out,k})} \tag{3}$$

2.2. Machine Learning Models

The different mathematical models (or black-box models) built to estimate the HLC of the analysed building are presented in this section. In this case, the three machine learning algorithms studied are XGBoost, SVR and MLP neural network.

2.2.1. Extreme Gradient Boosting

Gradient boosting is a meta-algorithm built on the basis of weak learners, like decision trees [19], with the aim of obtaining a strong ensemble learner [46,47]. Specifically, XGBoost is a scalable and efficient way to carry out the implementation of a gradient boosting algorithm. It is remarkable for its ease of implementation and its high accuracy in predictions [20]. This algorithm focuses on the idea of combining, in a final step, all the predictions made by a set of weak learners (*additive training strategy* [48]). Moreover, XGBoost is capable of simplifying the objective function with different combinations of predictive and regularization terms without loss in the optimal computational speed [19,20]. The specific learning process is summarized as follows [18,20]:

- An initial learner is fitted to the whole sample of inputs.
- A second model is fitted to the residuals of the first model to reduce its deficiencies.
- These two learning steps are repeated until a particular stop criterion is reached.
- The final predictions are obtained from the sum of the individual predictions of the learners used. The general function to obtain a prediction at step t, with additive training, is presented in Equation (4).

$$G_i^{(t)} = \sum_{j=1}^{t} G_j(\mathbf{x}_i) = G_i^{(t-1)} + G_t(\mathbf{x}_i) \tag{4}$$

where $G_t(\mathbf{x}_i)$ is the learner at step t, $G_i^{(t-1)}$ the prediction at step $t-1$ and $\mathbf{x}_i$ the input variable.

On the other hand, to prevent the problem of overfitting while maintaining an optimal computational speed, XGBoost takes into account Equation (5) to evaluate the suitability of the model [17,20]:

$$O^{(t)} = \sum_{j=1}^{n} L(\hat{\mathbf{y}}_i, \mathbf{y}_i) + \sum_{j=1}^{t} \Psi(G_i) \tag{5}$$

n being the number of observations, L a differentiable loss function and Ψ a regularization term that penalizes the complexity of the model [18].

XGBoost expands the loss function into the second order to be able to optimize the problem quickly. Furthermore, several specific techniques can reduce the possible overfitting of the algorithm [18]. One of them consists of, after each step of boosting, the weights recently added being scaled by a factor η (*shrinkage*). This process reduces the influence of the individual trees built, and therefore, the training will be slower and more efficient.

Lastly, XGBoost optimization requires the control of multiple parameters [19], and finding the best combination of them is important. In this analysis, the optimal combinations of parameters (*max_depth*, *min_child_weight, subsample, colsample_bytree* and *learning_rate*) are found through a k-fold cross-validation method ($k = 10$) [49] carried out on the different validation samples studied.

2.2.2. Support Vector Regression

SVR is a nonlinear kernel based regression model that focuses on finding the best regression hyperplane with the least structural risk in a high-dimensional feature space [25–27]. The SVR function is represented in Equation (6):

$$g(\mathbf{x}) = \mathbf{w}^T \delta(\mathbf{x}) + \mathbf{b} \tag{6}$$

$\delta(x)$ being a nonlinear mapping that connects the input space to the feature space, b a bias term and w the weight coefficient vector. In this case, both w and b are estimated by resolving the following optimization problem [26,28]:

$$\begin{aligned} \underset{\mathbf{w},\mathbf{b}}{\text{minimize}} \quad & \frac{1}{2} \parallel \mathbf{w} \parallel^2 + C \sum_{j=1}^{l} (\tau_j + \tau_j^*) \\ \text{subject to} \quad & \mathbf{y}_j - g(\mathbf{x}_j) \leq \epsilon + \tau_j \\ & g(\mathbf{x}_j) - \mathbf{y}_j \leq \epsilon + \tau_j^* \\ & \tau_j, \tau_j^* \geq 0 \end{aligned} \tag{7}$$

where the constant $C > 0$ represents the trade-off between training error and model complexity, ϵ corresponds to a threshold value and l is the number of training patterns. As Huang et al. [28] and Vrablecová et al. [29] showed, once having resolved the optimization problem (Equation (7)) and taking the Lagrangian, the model solution can be reached with its dual representation (see Equation (8)):

$$g(\mathbf{x}) = \sum_{j=1}^{l} (\alpha_j^* - \alpha_j) K(\mathbf{x}, \mathbf{x}_j) + \mathbf{b} \tag{8}$$

where α_j, α_j^* are the Lagrangian multipliers ($\neq 0$) and the solution for the dual problem, b the bias term and $K(x_j, x)$ the kernel function based on the inner product $\langle \delta(x_j), \delta(x) \rangle$.

Specifically, the model optimization for each of the validation samples studied is carried out through the tuning of certain parameters. In this case, the selected parameters were penalty C and term ϵ [24,29]. With a cross-validation process (k-fold = 10) [49], different values of these parameters were evaluated, and the best adjustment was obtained depending on the validation sample considered.

2.2.3. Multi-Layer Perceptron Neural Network

MLP is an ANN characterized by having several layers [32,40,50]:

- An input layer (first layer), where the inputs are introduced.
- An output layer (last layer), where the results obtained by the trained model are given.
- Hidden layers (intermediate layers) positioned between the previous ones. They can be zero, one or more.

The specific neural network built in this study is composed of five layers (one input layer, one output layer and three hidden layers). The internal structure of neurons in each layer of the network is as follows: 100-100-100-50-1. This architecture is selected after a k-fold cross-validation (k = 10) study [51] in which the grid of different hidden layer options was between zero and four [52,53]. On the other hand, in relation to the number of neurons in the network layers, the different options, as recommended by Vujicic et al. [52] and Doukim et al. [54], take into account the size of the sample and the number of inputs and outputs.

Moreover, due to the complexity of the problem, this grid of values was completed with more complex options such as: 50, 100, 200 and 400.

The neural network training was developed via backward propagation; errors were propagated and corrected through the network [55]. In this way, the real outputs must be known. In addition, the training process was based on updating the weights with an average update (batch-learning), which was obtained by introducing all the patterns in the input file (an epoch) and accumulating the weight updates [56,57]. The stop criterion used to avoid overfitting problems was the cross-validation due to its effectiveness in stopping with the best model generalization [58,59]. The training will stop when the performance of the neural network, measured by the Mean Squared Error (MSE) and evaluated with a small part of the whole sample (test sample), stagnates or starts to decrease. Each MLP trained in this study used the Rectified Linear Unit (ReLU) as the activation function [60], a normal kernel initializer and the Adaptive Moment Estimation (Adam) optimizer [61]. Lastly, the batch size with which the neural networks were trained was equal to 64, and the limit to the epochs in which the model fit did not improve was 100. Further information about the options and variations of the MLP training process can be found in [62,63].

2.3. *Case Study Data Acquisition*

2.3.1. Description of the Building

The methodology proposed was tested on a public library of the Faculty of Marine Sciences at the University of Vigo (see Figure 1) that is located in the northwest of Spain. The building has three floors that are interconnected, and it is completely monitored and has been used for research purposes many times. Therefore, the building, its HVAC system and its data acquisition system were further and more deeply described in other articles such as Cacabelos et al. [64,65], Fernández et al. [66] and Martínez et al. [40]. The building envelope has a large, south-facing window (see Figure 1a), and its enclosures are made of different concrete and insulation material. Table 1 shows the properties and composition of each material layer of the walls, floor and roof.

(a) (b)

Figure 1. Appearance of the analysed building: (**a**) exterior and (**b**) interior.

Table 1. Composition and properties of the enclosures of the analysed building.

Layer (Indoor-Outdoor)	Material	Thickness (cm)	Properties			
			λ (W/m · K)	c (kJ/kg · K)	ρ (kg/m^3)	R (h · m^2 · K/kJ)
EXTERIOR WALLS						
1	Plasterboard	2	0.11	1	900	-
2	Extruded polystyrene	4	0.03	1	31	-
3	Mineral wool	6	0.04	1	-	-
4	Air	4	-	-	-	0.05
5	Concrete	25	1.15	1	1800	-
INTERIOR FLOOR						
1	Extruded polystyrene	4	0.03	1	31	-
2	Greenket	2	0.10	1.6	300	-
3	Common concrete	8	1.3	1	2000	-
4	Lightweight concrete	6	0.34	1.1	600	-
5	Concrete block	2	1.32	1	1330	-
SLAB						
1	Extruded polystyrene	4	0.03	1	31	-
2	Greenket	2	0.10	1.6	300	-
3	Common concrete	14	1.3	1	2000	-
4	Lightweight concrete	6	0.34	1.1	600	-
5	Concrete block	2	1.32	1	1330	-
ROOF						
1	Plasterboard	2	0.11	1	900	-
2	Extruded polystyrene	3	0.03	1	31	-
3	Com on concrete	10	1.3	1	2000	-
4	Reinforced concrete	36	2.30	1	2400	-
5	Air	5	-	-	-	0.044

2.3.2. Pre-Processing Data

This analysis is focused on the estimation of the HLC of a building. The data available in this study to train the machine learning models were hourly observations between March 2016 and December 2017 of four variables. The variables considered were two describing the thermal behaviour of the building (thermal demand and indoor temperatures) and two describing the climate conditions (outdoor temperature and solar radiation). In addition, to find significant hours when the heating worked normally, only the hours with a thermal demand of more than 5 kW were selected (n = 5727). In this case, it was not necessary to have a continuous sample because the training did not take into account time lags in the explanatory variables. Therefore, three time variables (hour of the year, day of the week, and hour of the day) were introduced into the models to provide more information and thus to improve the training. The variables that were used as model inputs are presented in Figure 2.

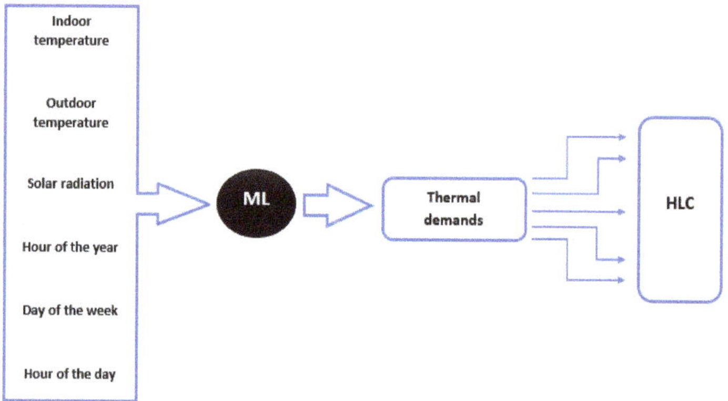

Figure 2. Summary of the training and prediction process. On the left, the input variables of the models. On the right, the objective estimations of the Heat Loss Coefficient (HLC).

The model fit was carried out through a cross-validation process in which the whole sample was divided consecutively into two individual samples: training and testing. In the search for the optimal model and the selection of hyperparameters, the models were tested with different partitions (or subsamples) of the whole training sample (*k*-fold = 10) [49]. In addition, the models, after being trained and tested with 10 different test samples, were validated with three independent samples to prove their efficiency in estimating the HLC of the building:

- Sample 1: 15/12/2018 07:00 – 17/12/2018 07:00
- Sample 2: 19/01/2019 07:00 – 21/01/2019 07:00
- Sample 3: 09/03/2019 07:00 – 11/03/2019 07:00

These three samples had to comply with certain restrictions due to the specification of the HLC formula (see Section 2.1) and because of the singularities of the monitoring of the building studied (see Section 2.3.1). The restrictions for considering a period suitable for the calculation of the HLC are [8,44]:

- **Weekend:** Due to the non-availability of occupation data, the period must be on the weekend when there was no occupancy in the building.
- **Cool period:** The average difference between indoor and outdoor temperature must be 10 K or more.
- **Cloudy period:** Solar radiation must be low: gains from solar radiation equal to or below 10% of the thermal demand.
- **Temperature stability:** The average between indoor and outdoor temperature must be similar at the beginning and end of the period to ensure steady state conditions.

2.3.3. Validation and Error Assessment

The Coefficient of Variation of the Root Mean Squared Error (CV(RMSE)) and the Normalized Mean Biased Error (NMBE) were the error measures calculated to evaluate the accuracy of the models presented:

$$\text{CV(RMSE)} = 100 \times \frac{\sqrt{\sum_{i=1}^{N}(y_i - \hat{y}_i)^2/N}}{\bar{y}} \tag{9}$$

$$\text{NMBE} = 100 \times \frac{\sum_{i=1}^{N}(y_i - \hat{y}_i)}{\sum_{i=1}^{N}(y_i)} \tag{10}$$

Both measurements were used to compare the performance of the different models in the three validation samples. In addition, the CV(RMSE) was also taken into account to select the best trials of each model (with the lowest error) and, then, represented graphically in the Results Section. These measures have been used and their efficiency has been proven in similar studies [42,67,68].

3. Results and Discussion

A methodology, based on thermal demands' predictions made with black-box models, to estimate the HLC of a building is developed in this paper. In particular, the building under study was the public library of the Faculty of Marine Sciences at the University of Vigo. The available data were hourly observations of the variables presented in Figure 2 from March 2016 to December 2017; only taking into account the hours with a significative thermal demand ($\geq$ 5 kW). In this analysis, three specific weekends were considered to study the performance of the models estimating the HLC through heating demand predictions. A comparison between the accuracy of each ML model analysed is presented in the following sections.

Section 3.1 presents the models performance in heating demand predictions and Section 3.2 a similar analysis for the HLC estimations. While in the thermal demand analysis, CV(RMSE) (Equation (9)) and NMBE (Equation (10)) were calculated for each of the models, in the HLC section, the errors were calculated based on an absolute variation rate. Furthermore, to represent the average performance of the different models, the predictions of the three validation samples were repeated 10 times (varying the subsamples where the model was tested). Thus, the numerical results shown are average errors obtained through the 10 trials, as well as their standard deviations. Finally, all figures shown in the following sections were created with the Python programming language [69].

3.1. Thermal Demand Analysis

The results of the heating demand predictions for the analysed building in the three validation samples are presented in this section in Figure 3 and Table 2. These predictions are important as they will be used to obtain the subsequent HLC estimations (see Section 3.2). Furthermore, in Figure 3, each of the algorithms is represented by the prediction that obtains the lowest CV(RMSE) among the 10 trials (best scenario).

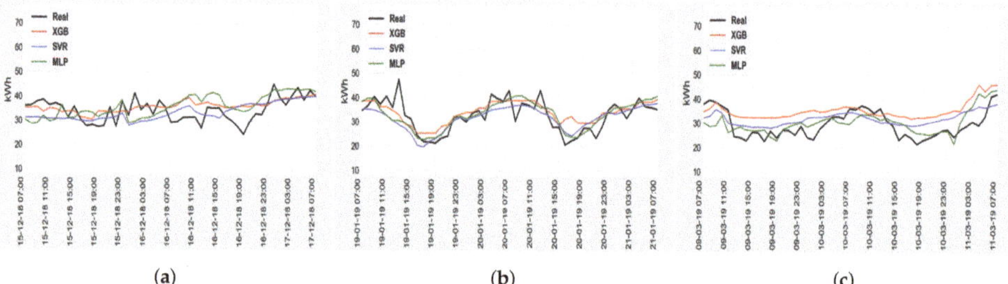

Figure 3. Results of thermal demand predictions in all samples studied: (**a**) Sample 1; (**b**) Sample 2; (**c**) Sample 3. Each machine learning model is represented by the prediction curve with the lowest CV(RMSE) among the 10 experiment trials.

In the case of Sample 1 (Figure 3a) and taking into account the CV(RMSE) results, the XGBoost model presents the lowest average error (18.57%). The MLP neural network and SVR model show a worse performance with an average error of 19.84% and 21.31%, respectively (see Table 2). Moreover, Table 2 shows that in all models, the variability of errors in the 10 trials was relatively low (below ± 4), but MLP is the one that shows the lowest dispersion (±2.41). On the other hand, regarding the NMBE results, the lowest average error was obtained by the MLP model (−5.49%), while the other algorithms only managed to obtain negative average errors higher than 10% (−12.52% for XGBoost and −11.63% for SVR). In addition, in this case, while the variability of the NMBE results of XGBoost and MLP are similar and around ± 5 (see Table 2), the SVR results show a greater dispersion (above ± 8). In general, all the models, in the best scenario, efficiently reproduced the real behaviour of the thermal demand of the building analysed (see Figure 3a). Although the XGBoost model obtained better results in relation to the CV(RMSE), the MLP predictions showed a better overall adjustment to reality by obtaining a similar average CV(RMSE) and an average NMBE much lower than XGBoost.

In Sample 2 (Figure 3b), considering the CV(RMSE) results, the XGBoost algorithm and MLP neural network show a similar average performance. MLP obtained an average error of 17.38% and XGBoost 17.43%. In this situation, the SVR model shows better results than the other models with an average error of 16.61% (see Table 2). Additionally, while SVR and MLP present a similar error dispersion around ± 2, the XGBoost model is the one that presents the highest variability (above ± 4). With respect to the NMBE results, the SVR model again obtained the lowest average error (2.99%). The MLP neural network and XGBoost algorithm, on the other hand, present higher and negative average error (−4.01% and −8.40%, respectively). However, as shown in Table 2, the SVR model is the one that showed the greatest dispersion in the NMBE results (almost ± 7), even though the other models also showed an important variability. In Figure 3b, it is demonstrated that each of the algorithms, taking into account the best scenario, is capable of replicating the reality except for certain peaks. However, more specifically, the model that yielded the best average performance was SVR.

Regarding Sample 3 (see Figure 3c), the three models studied obtained average CV(RMSE) values above 20%. While the SVR algorithm and MLP neural network showed an average CV(RMSE) of 21.60% and 21.54%, respectively, XGBoost model obtained an average error around 29% (see Table 2). Furthermore, taking into account that the variabilities presented were not high, the MLP neural network obtained the least variable results (±2.33) and XGBoost the most variable (±3.81). In the case of the NMBE results, the MLP model stands out from the rest (see Table 2). It shows the lowest average of NMBE by far (0.25%). The average results of the other algorithms were higher than 10% and negative (−13.60% for SVR and −23.05% for XGBoost). On the other hand, the dispersion of the NMBE values among the 10 trials of this sample is high (above 4.5) in all the models, and the MLP neural network was the model with the greatest error dispersion (±6.19). In addition, as in the other validation samples, Figure 3c shows that the built models, in the best case scenario, are very close to the real values.

Table 2. Numerical results of the thermal demand predictions for all validation samples. The mean of the CV(RMSE) and the Normalized Mean Biased Error (NMBE) obtained through the 10 trials, besides the standard deviation (SD) of each of them, are presented.

Model	Sample 1				Sample 2				Sample 3			
	CV(RMSE) (%)	SD	NMBE (%)	SD	CV(RMSE) (%)	SD	NMBE (%)	SD	CV(RMSE) (%)	SD	NMBE (%)	SD
XGBoost	18.570	3.940	−12.517	5.224	17.434	4.203	−8.404	4.759	29.143	3.808	−23.052	4.473
SVR	21.312	3.710	−11.627	8.460	16.607	2.417	2.987	6.819	21.596	3.581	−13.600	5.049
MLP	19.839	2.409	−5.491	5.693	17.381	2.301	−4.012	3.061	21.543	2.331	0.253	6.189

Lastly, it is demonstrated that all the models presented, in general, are capable of recreating reality (see Figure 3), obtaining optimal errors in all the samples studied (see Table 2). Concretely, they normally predict above the real values of the heating demand of the building analysed (see the negative NMBE results in Table 2). Furthermore, encompassing all the results, the model that performed best was the MLP neural network. In terms of CV(RMSE), it always yielded one of the best results, and in terms of NBME, it obtained much better results than the other algorithms in two of the three validation samples.

3.2. HLC Estimation Analysis

The specific characteristics of each of the validation samples studied in this work are analysed in Table 3 and presented in Figure 4. Moreover, the results of the HLC estimations, based on the previous heating demand predictions, for each of the three validation samples are shown in this section in Table 4. Although estimated from the heating demand predictions during a weekend, the HLC is represented as a single number (see Equation (3)). As in the preceding section, the average performance of the different algorithms among the 10 trials is summarized in Table 4.

Table 3. Thermal conditions of the three validation samples. These represent the mean difference between indoor and outdoor temperatures, the average heating demand ($\overline{Q}$), the solar radiation gain (Rad) over the thermal demand and the average temperatures (indoor and outdoor) at the start ($\overline{T^{\circ}}_{initial}$) and at the end of the period ($\overline{T^{\circ}}_{final}$). HLC, Heat Loss Coefficient.

	$HLC_{calculated}$ (kW/K)	$\overline{T_{int} - T_{out}}$ (K)	$\overline{Q}$ (kW)	Rad/Q (%)	$\overline{T^{\circ}}_{initial}$ (K)	$\overline{T^{\circ}}_{final}$ (K)
Sample 1	2.750	12.476	34.313	0.091	289.671	288.127
Sample 2	2.146	15.246	32.725	0.127	287.917	287.220
Sample 3	2.464	12.053	29.704	0.232	289.402	287.741

First, Table 3 shows that the three validation samples fulfilled the conditions necessary to efficiently measure the HLC of the analysed building (see Section 2.3.2). Moreover, each of the samples has different HLC values. The main different thermal conditions of the samples, which caused slightly different HLC values, are also summarized in Table 3 and presented in Figure 4. The highest HLC value was obtained in Sample 1 (2.75 kW/K) due to the fact that the weight of the radiation gains, in relation to the heating demands, was the lowest among the samples (0.09%). This is related to the fact that the average thermal demand throughout this period was the highest and is a numerator value in the HLC formula (see Equation (3)). In addition, the average difference between indoor and outdoor temperatures, as shown in Figure 4a, was one of the smallest (12.48 K). These were the main reasons for obtaining a greater HLC (see Table 3). On the other hand, Sample 2 is where the HLC value is the lowest (2.15 kW/K) because, as shown in Figure 4b, the average difference between indoor and outdoor temperatures was the largest (15.25 K). This value, which is in the denominator of the HLC formula, together with an average heating demand significantly lower than in Sample 1 (32.72 kW) reduces the calculated HLC value (see Table 3). Lastly, in Sample 3 (see Figure 4c), an intermediate value of the HLC (2.46 kW/K) that came from the lowest average difference between indoor and outdoor temperatures (12.05 K) and the lowest average thermal demand (29.70 kW) was obtained.

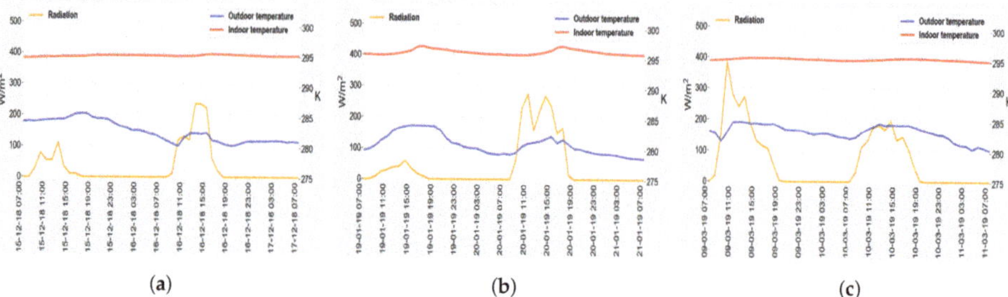

Figure 4. Summary of the thermal conditions of each of the validation samples: (**a**) Sample 1; (**b**) Sample 2; (**c**) Sample 3. Solar radiation and indoor and outdoor temperatures affecting the analysed building during the specific time periods are represented.

On the other hand, in the case of Sample 1, where the calculated HLC was 2.75 kW/K, the MLP neural network was the most accurate model (see Table 4). While this algorithm obtained an average absolute variation rate of 6.50%, the XGBoost and SVR models were only able to obtain an average absolute variation greater than 10% (12.52.% and 12.16%, respectively). Thus, the average HLC value estimated by MLP (2.90 kW/K) is the closest to the measured HLC value. Table 4 shows that all models obtained a higher average estimation than the calculated one (the same situation as in Section 3.1). Additionally, in relation to the variation rate dispersion, the MLP neural network was the one with the lowest variability among the errors (±4.5). However, in general, all models showed a significant high standard deviation.

Table 4. Numerical results of the HLC estimations for all validation samples. The values shown summarize the performance of each of the models through 10 repetitions of the experiment. The mean of the absolute variation rates (together with the standard deviation), the mean of the estimated HLC values and the calculated HLC are presented.

Model	Sample 1—$HLC_{calculated}$ = 2.750			Sample 2—$HLC_{calculated}$ = 2.146			Sample 3—$HLC_{calculated}$ = 2.464		
	$HLC_{estimated}$	% Variation	SD	$HLC_{estimated}$	% Variation	SD	$HLC_{estimated}$	% Variation	SD
XGBoost	3.094	12.517	5.224	2.327	8.404	4.759	3.032	23.052	4.473
SVR	3.070	12.158	7.677	2.082	5.680	4.813	2.800	13.600	5.049
MLP	2.901	6.505	4.500	2.232	4.083	2.966	2.458	4.966	3.702

In Sample 2 the calculated HLC was 2.15 kW/K, and in relation to the average absolute variation values, the most accurate model was the MLP neural network (4.08%), but close to the SVR model (5.68%). As in the thermal demand section, the XGBoost model performed worse with an average absolute variation rate of 8.40% (see Table 4). Therefore, the HLC values estimated by the SVR and MLP models are very close to the calculated HLC value; while the average value obtained by SVR is 2.08 kW/K, the average value estimated for MLP is 2.23 kW/K. With respect to the dispersion of variation rate data and, as in Sample 1,

Table 4 shows that the MLP neural network was the most stable model with a standard deviation below ±3. The other models presented values above ±4.

Regarding Sample 3, in which the measured HLC was 2.46 kW/K, the MLP neural network was again the model with the best average performance. While the XGBoost and SVR algorithms presented average absolute variation of 23.05% and 13.60%, respectively, the MLP model showed an average absolute variation of 4.97% (see Table 4). Therefore, the average HLC value from the MLP estimations (2.46 kW/K) was much closer to the calculated HLC value than those obtained by the other models. In addition, in this sample, all models obtained a high variability in their results: all standard deviations were higher than ±3.5. In this situation, the MLP neural network was the model with the lowest dispersion among its errors (±3.70).

Definitely, the HLC value that characterizes the studied building, calculated as the average of the presented results, was 2.45 ± 0.30 kW/K. Observing the results presented in Table 4 and taking into account the whole analysis (the results of the three validation samples), the model that presents the best average performance was the MLP neural network. It was the model that obtained the most stable results and had the lowest average absolute variation rate for all the samples analysed. In this way, it is demonstrated that it is possible to estimate the HLC of the analysed building with an absolute error around 4 or 6% if an MLP model is used to make the necessary thermal demand predictions. On the other hand, if an SVR algorithm is used, the error increases to 5-13%, and if the model is XGBoost, the errors vary between 8 and 23%.

4. Conclusions

A new methodology for estimating the HLC of a building is presented in this paper. It is based on the introduction of thermal demand predictions obtained with machine learning models in the HLC formula. This study focuses on the analysis, on the one hand, of monitored data on heating demands and indoor temperatures belonging to the Science Library of the University of Vigo. On the other hand, two meteorological variables (outdoor temperature and solar radiation) and three temporal variables (hour of the year, day of the week and hour of the day) are also taken into account. The aim of this paper is to show a methodology that allows the efficient estimation of the HLC value of a building without the need to control its heating demands (nor indoor temperatures if they are assumed to be constant). The search for the optimal methodology considers and compares three different machine learning models (XGBoost, SVR and MLP neural network). In addition, the performance of each one is evaluated and analysed both through its average accuracy in thermal demand predictions and its average accuracy in HLC estimations.

The research contribution of this work is the application of mathematical models to estimate the HLC of a building with low errors. In addition to reducing the necessity for monitoring, these models can be useful for detecting errors in measurements from sensors installed in the building. Moreover, the black-box models presented contributes with advantages compared to traditional research in building simulation. The use and application of the traditional building thermal simulation models are conditioned by the need for significant knowledge on a subject. In addition, these models need to control many different parameters related to the energy performance of a building. Nevertheless, machine learning models, which need less time for development than dynamic simulation methods, do not require specific prior knowledge and can be applied in numerous fields. The only important need for these models is the availability of a significant amount of data from which a behaviour pattern is extracted. Furthermore, the inputs introduced in the built models are variables typically monitored in buildings. Therefore, the methodology presented here is extractable to other studies and buildings.

The results obtained show that it is possible to efficiently estimate the HLC value of a building over specific time periods with black-box models. To this end, it is important to obtain thermal demand predictions, which are used as inputs in HLC estimations, with average errors lower than the values proposed for the calibrated models. The results also demonstrate that the MLP neural network is the algorithm with the best average performance in HLC estimation (more stability and higher average accuracy in all samples studied). The SVR model shows a close average behaviour, but XGBoost, except in Sample 1, presents a much worse performance than the other two. In the first validation sample, in which the measured HLC is 2.750 kW/K, only the MPL neural network is capable of obtaining an average absolute variation rate below 10% (6.50%). The SVR and XGBoost models obtain an average variation of 12.16% and 12.52%, respectively. On the other hand, in the second validation sample, where the calculated HLC was 2.146 kW/K, both the MLP and SVR models performed better than XGBoost. While the first two obtain an average absolute variation rate of 4.08% and 5.68%, respectively, XGBoost shows an average variation of 8.40%. Lastly, in the case of the third validation sample, with a measured HLC value of 2.464 kW/K, the MLP neural network yields an average absolute variation rate far from the other algorithms (4.97%). The SVR and XGBoost models are only able to yield an average absolute variation of 13.60% and 23.05%, respectively. Furthermore, regarding the dispersion in error data, MLP model is the one that shows the most stable results in all validation samples. The other algorithms present similar variability between them, but higher than the one obtained by the MLP model. Definitely, taking into account all the results presented, the HLC value of the analysed building is 2.45 ± 0.30 kW/K.

From an energy point of view, the conclusion is that efficient predictions related to the thermal conditions of a building and, in addition, made by machine learning models can be used to efficiently estimate its HLC. This is demonstrated in this paper with three different validation samples that fulfil the necessary conditions to calculate the HLC. The most accurate model, which in this case is the MLP neural network, is able to estimate the HLC of the analysed building with an average absolute variation rate of around 5% and with a standard deviation of around ± 3. On the other hand, the main limitation of this research is the many restrictions on finding suitable time periods for calculating the HLC. In this case, the unavailability of occupation data means that only weekends are studied. For this reason, some possible future lines of research are similar analyses considering more data such as occupation or extending the study to more general situations.

Author Contributions: Conceptualization, L.F.-G. and M.M.-C.; methodology, M.M.-C.; software, M.M.-C.; validation, L.F.-G., E.G.-Á. and S.M.-M.; formal analysis, L.F.-G., J.M.-T. and S.M.-M.; investigation, M.M.-C., L.F.-G. and S.M.-M.; resources, E.G.-Á.; data curation, M.M.-C. and J.M.-T.; writing, original draft preparation, M.M.-C. and L.F.-G.; writing, review and editing, L.F.-G., J.M.-T., E.G.-Á. and S.M.-M.; visualization, M.M.-C.; supervision, E.G.-Á., L.F.-G. and J.M.-T.; project administration, E.G.-Á.; funding acquisition, E.G.-Á. All authors read and agreed to the published version of the manuscript.

Funding: This research was funded by the Spanish Government (Science, Innovation and Universities Ministry) under the project RTI2018-096296-B-C21.

Acknowledgments: This research was supported by the Spanish Government (Science, Innovation and Universities Ministry) under the project RTI2018-096296-B-C21.

Conflicts of Interest: The authors declare no conflict of interest.

References

1. *IEA EBC Annex 75: 5th Expert Meeting*; Energy in Buildings and Communities Programme (EBC): Venice, Italy, 2020.
2. *Directive 2010/31/EU of the European Parliament and of the Council of 19 May 2010 on the Energy Performance of Buildings*; Official Journal of the European Union: Brussels, Belgium, 2010; Volume 153, pp. 13–35.
3. *Directive 2012/27/EU of the European Parliament and of the Council of 25 October 2012 on Energy Efficiency, Amending Directives 2009/125/EC and 2010/30/EU and Repealing Directives 2004/8/EC and 2006/32/EC Text with EEA Relevance*; European Commission: Brussels, Belgium, 2012; Volume 315, pp. 1–56.

4. *Directive 2018/844/EU of the European Parliament and of the Council of 30 May 2018 Amending Directive 2010/31/EU on the Energy Performance of Buildings and Directive 2012/27/EU on Energy Efficiency*; European Commission: Luxembourg, 2018; Volume 156, pp. 75–91.
5. Gatt, D.; Yousif, C.; Cellura, M.; Camilleri, L.; Guarino, F. Assessment of building energy modelling studies to meet the requirements of the new Energy Performance of Buildings Directive. *Renew. Sustain. Energy Rev.* **2020**, *127*, 109886. [CrossRef]
6. Yan, D.; Hong, T.; Dong, B.; Mahdavi, A.; D'Oca, S.; Gaetani, I.; Feng, X. IEA EBC Annex 66: Definition and simulation of occupant behavior in buildings. *Energy Build.* **2017**, *156*, 258–270. [CrossRef]
7. De Wilde, P. The gap between predicted and measured energy performance of buildings: A framework for investigation. *Autom. Constr.* **2014**, *41*, 40–49. [CrossRef]
8. Uriarte, I.; Erkoreka, A.; Giraldo-Soto, C.; Martin, K.; Uriarte, A.; Eguia, P. Mathematical development of an average method for estimating the reduction of the Heat Loss Coefficient of an energetically retrofitted occupied office building. *Energy Build.* **2019**, *192*, 101–122. [CrossRef]
9. Givoni, B. Well Tempered and Illuminated Interiors. In *Passive and Low Energy Ecotechniques*; Bowen, A., Ed.; Pergamon: Oxford, UK, 1985; pp. 210–225. [CrossRef]
10. Maile, T.; Fischer, M.; Bazjanac, V. Building Energy Performance Simulation Tools-a Life-Cycle and Interoperable Perspective. In *Center for Integrated Facility Engineering (CIFE) Working Paper*; CIFE: Stanford, CA, USA, 2007; Volume 107.
11. Eguía Oller, P.; Alonso Rodríguez, J.; Saavedra González, A.; Arce Fariña, E.; Granada Álvarez, E. Improving transient thermal simulations of single dwellings using interpolated weather data. *Energy Build.* **2017**, *135*, 212–224. [CrossRef]
12. Lü, X.; Lu, T.; Kibert, C.J.; Viljanen, M. Modeling and forecasting energy consumption for heterogeneous buildings using a physical–statistical approach. *Appl. Energy* **2015**, *144*, 261–275. [CrossRef]
13. Li, X.; Wen, J. Review of building energy modelling for control and operation. *Renew. Sustain. Energy Rev.* **2014**, *37*, 517–537. [CrossRef]
14. Helm, M.; Swiergosz, A.; Haeberle, H.; Karnuta, J.; Schaffer, J.; Krebs, V.; Spitzer, A.; Ramkumar, P. Machine Learning and Artificial Intelligence: Definitions, Applications, and Future Directions. *Curr. Rev. Musculoskelet. Med.* **2020**, *13*. [CrossRef] [PubMed]
15. Murdoch, W.J.; Singh, C.; Kumbier, K.; Abbasi-Asl, R.; Yu, B. Interpretable machine learning: definitions, methods, and applications. *arXiv* **2019**, arxiv:abs/1901.04592.
16. Pesantez-Narvaez, J.; Guillen, M.; Alcañiz, M. Predicting Motor Insurance Claims Using Telematics Data—XGBoost versus Logistic Regression. *Risks* **2019**, *7*, 10. [CrossRef]
17. Babajide Mustapha, I.; Saeed, F. Bioactive Molecule Prediction Using Extreme Gradient Boosting. *Molecules* **2016**, *21*, 983. [CrossRef] [PubMed]
18. Wang, H.; Liu, C.; Deng, L. Enhanced Prediction of Hot Spots at Protein-Protein Interfaces Using Extreme Gradient Boosting. *Sci. Rep.* **2018**, *8*. [CrossRef] [PubMed]
19. Carmona, P.; Climent, F.; Momparler, A. Predicting failure in the U.S. banking sector: An extreme gradient boosting approach. *Int. Rev. Econ. Financ.* **2019**, *61*, 304–323. [CrossRef]
20. Fan, J.; Wang, X.; Wu, L.; Zhou, H.; Zhang, F.; Yu, X.; Lu, X.; Xiang, Y. Comparison of Support Vector Machine and Extreme Gradient Boosting for predicting daily global solar radiation using temperature and precipitation in humid subtropical climates: A case study in China. *Energy Convers. Manag.* **2018**, *164*, 102–111. [CrossRef]
21. Guo, R.; Zhao, Z.; Wang, T.; Liu, G.; Zhao, J.; Gao, D. Degradation State Recognition of Piston Pump Based on ICEEMDAN and XGBoost. *Appl. Sci.* **2020**, *10*, 6593. [CrossRef]
22. Mo, H.; Sun, H.; Liu, J.; Wei, S. Developing window behavior models for residential buildings using XGBoost algorithm. *Energy Build.* **2019**, *205*, 109564. [CrossRef]
23. Fan, C.; Xiao, F.; Zhao, Y. A short-term building cooling load prediction method using deep learning algorithms. *Appl. Energy* **2017**, *195*, 222–233. [CrossRef]
24. Chen, K.Y.; Wang, C.H. Support vector regression with genetic algorithms in forecasting tourism demand. *Tour. Manag.* **2007**, *28*, 215–226. [CrossRef]

25. Zhong, H.; Wang, J.; Jia, H.; Mu, Y.; Lv, S. Vector field-based support vector regression for building energy consumption prediction. *Appl. Energy* **2019**, *242*, 403–414. [CrossRef]
26. Kazem, A.; Sharifi, E.; Hussain, F.K.; Saberi, M.; Hussain, O.K. Support vector regression with chaos-based firefly algorithm for stock market price forecasting. *Appl. Soft Comput.* **2013**, *13*, 947–958. [CrossRef]
27. Khelif, R.; Chebel-Morello, B.; Malinowski, S.; Laajili, E.; Fnaiech, F.; Zerhouni, N. Direct Remaining Useful Life Estimation Based on Support Vector Regression. *IEEE Trans. Ind. Electron.* **2017**, *64*, 2276–2285. [CrossRef]
28. Huang, K.; Guo, Y.F.; Tseng, M.L.; Wu, K.J.; Li, Z.G. A Novel Health Factor to Predict the Battery's State-of-Health Using a Support Vector Machine Approach. *Appl. Sci.* **2018**, *8*, 1803. [CrossRef]
29. Vrablecová, P.; Bou Ezzeddine, A.; Rozinajová, V.; Šárik, S.; Sangaiah, A.K. Smart grid load forecasting using online support vector regression. *Comput. Electr. Eng.* **2018**, *65*, 102–117. [CrossRef]
30. Paudel, S.; Nguyen, P.; Kling, W.; Elmitri, M.; Lacarrière, B.; Corre, O. Support Vector Machine in Prediction of Building Energy Demand Using Pseudo Dynamic Approach. *arXiv* **2015**, arxiv:abs/1507.05019.
31. Jiang, W.; He, G.; Long, T.; Ni, Y.; Liu, H.; Peng, Y.; Lv, K.; Wang, G. Multilayer Perceptron Neural Network for Surface Water Extraction in Landsat 8 OLI Satellite Images. *Remote Sens.* **2018**, *10*, 755. [CrossRef]
32. Azorin-Molina, C.; Ali, Z.; Hussain, I.; Faisal, M.; Nazir, H.M.; Hussain, T.; Shad, M.Y.; Mohamd Shoukry, A.; Hussain Gani, S. Forecasting Drought Using Multilayer Perceptron Artificial Neural Network Model. *Adv. Meteorol.* **2017**, *2017*, 5681308. [CrossRef]
33. Taki, M.; Ajabshirchi, Y.; Ranjbar, S.F.; Rohani, A.; Matloobi, M. Heat transfer and MLP neural network models to predict inside environment variables and energy lost in a semi-solar greenhouse. *Energy Build.* **2016**, *110*, 314–329. [CrossRef]
34. Iglesias, C.; Anjos, O.; Martínez, J.; Pereira, H.; Taboada, J. Prediction of tension properties of cork from its physical properties using neural networks. *Eur. J. Wood Wood Prod.* **2015**, *73*, 347–356. [CrossRef]
35. Anjos, O.; Iglesias, C.; Peres, F.; Martínez, J.; García, A.; Taboada, J. Neural networks applied to discriminate botanical origin of honeys. *Food Chem.* **2015**, *175*, 128–136. [CrossRef]
36. Chen, Y.; Song, L.; Liu, Y.; Yang, L.; Li, D. A Review of the Artificial Neural Network Models for Water Quality Prediction. *Appl. Sci.* **2020**, *10*, 5776. [CrossRef]
37. Kang, Y.; Lv, W.; He, J.; Ding, X. Remote Sensing of Time-Varying Tidal Flat Topography, Jiangsu Coast, China, Based on the Waterline Method and an Artificial Neural Network Model. *Appl. Sci.* **2020**, *10*, 3645. [CrossRef]
38. Chae, Y.T.; Horesh, R.; Hwang, Y.; Lee, Y. Artificial neural network model for forecasting sub-hourly electricity usage in commercial buildings. *Energy Build.* **2016**, *111*, 184–194. [CrossRef]
39. Kusiak, A.; Li, M.; Zhang, Z. A data-driven approach for steam load prediction in buildings. *Appl. Energy* **2010**, *87*, 925–933. [CrossRef]
40. Martínez Comesaña, M.; Febrero-Garrido, L.; Troncoso-Pastoriza, F.; Martínez-Torres, J. Prediction of Building's Thermal Performance Using LSTM and MLP Neural Networks. *Appl. Sci.* **2020**, *10*, 7439. [CrossRef]
41. Ruiz, G.R.; Bandera, C.F. Validation of Calibrated Energy Models: Common Errors. *Energies* **2017**, *10*, 1587. [CrossRef]
42. Hong, T.; Kim, J.; Jeong, J.; Lee, M.; Ji, C. Automatic calibration model of a building energy simulation using optimization algorithm. *Energy Procedia* **2017**, *105*, 3698–3704. [CrossRef]
43. Butler, D.; Dengel, A. *Review of Co-Heating Test Methodologies: Primary Research*; NHBC Foundation: Milton Keynes, UK, 2013.
44. Uriarte, I.; Erkoreka, A.; Eguia, P.; Granada, E.; Martin-Escudero, K. Estimation of the Heat Loss Coefficient of Two Occupied Residential Buildings through an Average Method. *Energies* **2020**, *13*, 5724. [CrossRef]
45. Duffie, J.; Beckman, W. *Solar Engineering of Thermal Processes*, 4th ed.; John Wiley and Sons: Hoboken, NJ, USA, 2013. [CrossRef]
46. Friedman, J.H. Stochastic gradient boosting. *Comput. Stat. Data Anal.* **2002**, *38*, 367–378. [CrossRef]
47. Touzani, S.; Granderson, J.; Fernandes, S. Gradient boosting machine for modelling the energy consumption of commercial buildings. *Energy Build.* **2018**, *158*, 1533–1543. [CrossRef]

48. Priscilla, C.V.; Prabha, D.P. Influence of Optimizing XGBoost to handle Class Imbalance in Credit Card Fraud Detection. In Proceedings of the 2020 Third International Conference on Smart Systems and Inventive Technology (ICSSIT), Tirunelveli, India, 2020; pp. 1309–1315. [CrossRef]
49. Xiong, Z.; Cui, Y.; Liu, Z.; Zhao, Y.; Hu, M.; Hu, J. Evaluating explorative prediction power of machine learning algorithms for materials discovery using k-fold forward cross-validation. *Comput. Mater. Sci.* **2020**, *171*, 109203. [CrossRef]
50. Pham, B.T.; Nguyen, M.D.; Bui, K.T.T.; Prakash, I.; Chapi, K.; Bui, D.T. A novel artificial intelligence approach based on Multi-layer Perceptron Neural Network and Biogeography-based Optimization for predicting coefficient of consolidation of soil. *CATENA* **2019**, *173*, 302–311. [CrossRef]
51. Sheela, K.; Deepa, S.N. Review on Methods to Fix Number of Hidden Neurons in Neural Networks. *Math. Probl. Eng.* **2013**, *2013*. [CrossRef]
52. Vujicic, T.; Matijević, T.; Ljucovic, J.; Balota, A.; Sevarac, Z. Comparative Analysis of Methods for Determining Number of Hidden Neurons in Artificial Neural Network. In *Central European Conference on Information and Intelligent Systems*; Faculty of Organization and Informatics Varazdin: Varaždin, Croatia, 2016.
53. Panchal, G.; Ganatra, A.; Kosta, Y.; Panchal, D. Behaviour Analysis of Multilayer Perceptrons with Multiple Hidden Neurons and Hidden Layers. *Int. J. Comput. Theory Eng.* **2011**, *3*, 332–337. [CrossRef]
54. Doukim, C.; Dargham, J.; Chekima, A. Finding the number of hidden neurons for an MLP neural network using coarse to fine search technique. In Proceedings of the 10th International Conference on Information Science, Signal Processing and their Applications (ISSPA 2010), Kuala Lumpur, Malaysia, 10–13 May 2010; pp. 606–609.
55. Liu, Y.; Liu, S.; Wang, Y.; Lombardi, F.; Han, J. A Stochastic Computational Multi-Layer Perceptron with Backward Propagation. *IEEE Trans. Comput.* **2018**, *67*, 1273–1286. [CrossRef]
56. Guresen, E.; Kayakutlu, G.; Daim, T.U. Using artificial neural network models in stock market index prediction. *Expert Syst. Appl.* **2011**, *38*, 10389–10397. [CrossRef]
57. Smith, L.N. A disciplined approach to neural network hyper-parameters: Part 1-learning rate, batch size, momentum, and weight decay. *arXiv* **2018**, arxiv:abs/1803.09820.
58. Li, M.; Soltanolkotabi, M.; Oymak, S. Gradient Descent with Early Stopping is Provably Robust to Label Noise for Overparameterized Neural Networks. In Proceedings of the Machine Learning Research (PMLR), Palermo, Italy, 2020; Volume 108, pp. 4313–4324.
59. Barrow, D.K.; Crone, S.F. Cross-validation aggregation for combining autoregressive neural network forecasts. *Int. J. Forecast.* **2016**, *32*, 1120–1137. [CrossRef]
60. Eckle, K.; Schmidt-Hieber, J. A comparison of deep networks with ReLU activation function and linear spline-type methods. *Neural Netw.* **2019**, *110*, 232–242. [CrossRef]
61. Bock, S.; Weiß, M. A Proof of Local Convergence for the Adam Optimizer. In Proceedings of the 2019 International Joint Conference on Neural Networks (IJCNN), Budapest, Hungary, 14–19 July 2019; pp. 1–8. [CrossRef]
62. Nakama, T. Theoretical analysis of batch and on-line training for gradient descent learning in neural networks. *Neurocomputing* **2009**, *73*, 151–159. [CrossRef]
63. Devarakonda, A.; Naumov, M.; Garland, M. AdaBatch: Adaptive Batch Sizes for Training Deep Neural Networks. *arXiv* **2017**, arXiv:1712.02029.
64. Cacabelos, A.; Eguía, P.; Míguez, J.L.; Granada, E.; Arce, M.E. Calibrated simulation of a public library HVAC system with a ground-source heat pump and a radiant floor using TRNSYS and GenOpt. *Energy Build.* **2015**, *108*, 114–126. [CrossRef]
65. Cacabelos, A.; Eguía, P.; Febrero, L.; Granada, E. Development of a new multi-stage building energy model calibration methodology and validation in a public library. *Energy Build.* **2017**, *146*, 182–199. [CrossRef]
66. Fernandez Rodríguez, M.; Eguía, P.; Granada, E.; Febrero Garrido, L. Sensitivity analysis of a vertical geothermal heat exchanger dynamic simulation: Calibration and error determination. *Geothermics* **2017**, *70*, 249–259. [CrossRef]
67. Kuo, P.H.; Huang, C.J. A High Precision Artificial Neural Networks Model for Short-Term Energy Load Forecasting. *Energies* **2018**, *11*, 213. [CrossRef]

68. Martínez, S.; Eguía, P.; Granada, E.; Moazami, A.; Hamdy, M. A performance comparison of multi-objective optimization-based approaches for calibrating white-box building energy models. *Energy Build.* **2020**, *216*, 109942. [CrossRef]
69. Pilgrim, M.; Willison, S. *Dive Into Python 3*; Springer: Berlin, Germany, 2009; Volume 2.

Publisher's Note: MDPI stays neutral with regard to jurisdictional claims in published maps and institutional affiliations.

Article

Artificial Intelligence, Accelerated in Parallel Computing and Applied to Nonintrusive Appliance Load Monitoring for Residential Demand-Side Management in a Smart Grid: A Comparative Study

Yu-Chen Hu [1], Yu-Hsiu Lin [2,*] and Chi-Hung Lin [2]

[1] Department of Computer Science and Information Management, Providence University, Taichung City 43301, Taiwan; ychu@pu.edu.tw

[2] Department of Electrical Engineering, Ming Chi University of Technology, New Taipei City 243303, Taiwan; m09128012@o365.mcut.edu.tw

* Correspondence: yhlin@mail.mcut.edu.tw

Received: 23 October 2020; Accepted: 13 November 2020; Published: 16 November 2020

Abstract: A smart grid is a promising use-case of AIoT (AI (artificial intelligence) across IoT (internet of things)) that enables bidirectional communication among utilities that arises with demand response (DR) schemes for demand-side management (DSM) and consumers that manage their power demands according to received DR signals. Disaggregating composite electric energy consumption data from a single minimal set of plug-panel current and voltage sensors installed at the electric panel in a practical field of interest, nonintrusive appliance load monitoring (NIALM), a cost-effective load disaggregation approach for (residential) DSM, is able to discern individual electrical appliances concerned without accessing each of them by individual plug-load power meters (smart plugs) deployed intrusively. The most common load disaggregation approaches are based on machine learning algorithms such as artificial neural networks, while approaches based on evolutionary computing, metaheuristic algorithms considered as global optimization and search techniques, have recently caught the attention of researchers. This paper presents a genetic algorithm, developed in consideration of parallel evolutionary computing, and aims to address NIALM, whereby load disaggregation from composite electric energy consumption data is declared as a combinatorial optimization problem and is solved by the algorithm. The algorithm is accelerated in parallel, as it would involve large amounts of NIALM data disaggregated through evolutionary computing, chromosomes, and/or evolutionary cycles to dominate its performance in load disaggregation and excessively cost its execution time. Moreover, the evolutionary computing implementation based on parallel computing, a feed-forward, multilayer artificial neural network that can learn from training data across all available workers of a parallel pool on a machine (in parallel computing) addresses the same NIALM/load disaggregation. Where, a comparative study is made in this paper. The presented methodology is experimentally validated by and applied on a publicly available reference dataset.

Keywords: artificial intelligence; artificial neural network; demand-side management; evolutionary computing; non-intrusive appliance load monitoring; parallel computing; smart grid; smart house

1. Introduction

Nonintrusive appliance load monitoring (NIALM), also called nonintrusive load monitoring, was first investigated by George W. Hart et al. [1] at the Massachusetts Institute of Technology with funding from the Electric Power Research Institute in the early 1980s. It has been considered as a cost-effective alternative against intrusive load monitoring approaches that involve the deployment of plug-load power meters (smart plugs) for individual concerned electrical appliances in a practical field of interest (it cannot be costed down for the realization of load management) [2], and has been developed for (residential) demand-side management (DSM) in a smart grid [3,4]. With DSM implemented in a smart grid, consumers have the opportunity to improve their awareness of what and when their monitored electrical appliances should operate for use in response to demand response (DR) schemes [5]. Thus, power utilities coming up with DR schemes plan to address ever-increasing electric demand in an optimal way [6].

NIALM, load disaggregation by estimating appliance-by-appliance energy consumption from composite electric energy consumption data, is becoming mature with the improvement of advanced metering infrastructure for DSM in a smart grid [7,8], where load disaggregation is performed with/applied on smart meter data.

NIALM can be built upon signal processing [9], machine learning [4,9–20], and deep learning [20–24]. Not much attention has been paid to evolutionary computing, wherein load disaggregation is considered as a combinatorial optimization problem [1]. As a result, in this paper, load disaggregation is declared as a combinatorial optimization problem, and is solved by a genetic algorithm (GA) accelerated in parallel computing. GA used to solve NIALM that is declared as a combinatorial optimization problem would involve large amounts of electric energy consumption data gathered from smart meters deployed in practical fields of interest in a smart grid, population-based candidate solutions evaluated, and/or evolutionary cycles executed from iteration to iteration, which is a highly demanding task in terms of performance in load disaggregation and computational time in evolutionary computation. Hence, GA conducted in this paper for NIALM is parallelized during evolution. The GA-based NIALM in this paper is processed and accelerated in consideration of parallel computing. It has been experimentally validated by and applied on a publicly-available reference dataset considered in this paper. Besides the GA-based NIALM, a feed-forward, multilayer artificial neural network (ANN)-based NIALM against other different types of ANNs (deep NNs)-based NIALM addresses the same NIALM data from the considered publicly available reference dataset. A comparative study is made in this paper. ANNs used in this paper to address NIALM can learn from NIALM data across parallel workers on a machine supporting parallel computing.

The remainder of this paper is structured below. The investigated methodology is presented in Section 2. Section 3 gives the experimentation and results. Conclusions are drawn in Section 4, where future work is also anticipated.

2. Methodology

Figure 1 describes a basic (eventless) NIALM process consisting of (1) data acquisition: composite electric energy consumption data are measured by a single minimal set of plug-panel current and voltage sensors and digitized for further analysis of load disaggregation; (2) feature extraction: electrical features are extracted as feature data from digitized composite electric energy consumption data for concerned electrical appliances; and (3) load recognition: artificial intelligence (AI) is utilized to recognize extracted electrical features for concerned electrical appliances—their electric demand based on past trends can be predicted. The NIALM investigated in this paper is an eventless NIALM approach. AI conducted in the NIALM is parallel computing accelerated evolutionary computing. The basic NIALM problem can be formulated as follows:

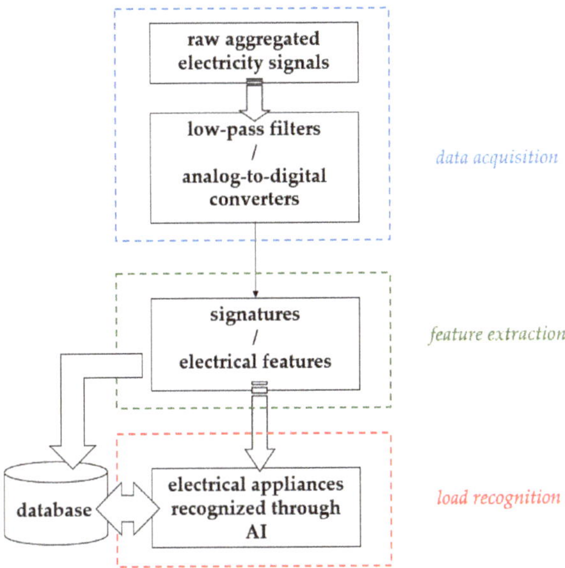

Figure 1. A basic NIALM process (profiling concerned electrical appliances and statistically computing base load should be done beforehand).

If (1) $P(t)$ stands for used composite power consumption acquired at time t and (2) $P_i(t)$ accounts for real power absorbed by concerned electrical appliance i (profiling concerned electrical appliances and base load is done in advance), the total power consumption, $P(t)$ (=$y(t)$), in an electric power distribution system can be expressed as Equation (1). In Equation (1), the summation of superimposed absorptions $\Sigma x_i(t)P_i(t)$ with an unmonitored base load $P_{base}(t)$ is made. In Equation (1), N is the total number of concerned electrical appliances and $x_i(t)$, a unit value from {0, 1}, is the on/off operational status of electrical appliance i at time t. NIALM is used to recognize the unknown (possible) operational status of concerned electrical appliances $x_i(t)|i = 1, 2, ..., N$ at time t when $P(t)$ (=$y(t)$) acquired and given with $P_{base}(t)$ is known: $[x_1(t), x_2(t), ..., x_i(t), ..., x_N(t)] = F(y(t))$ where F is a function that returns the best N estimates of $x_i(t)$ at time t for concerned individual N electrical appliances. F above can be any AI algorithm. In this paper, we conduct evolutionary computing, GA, to metaheuristically search for the optimal load combinations $x_i(t)$ of concerned electrical appliances when acquired total power consumption $P(t)$ with base load $P_{base}(t)$ is given at time t (in Equation (1), profiling concerned electrical appliances to get $P_i(t)$ and statistically computing base load for $P_{base}(t)$ from unmonitored electrical appliances is done prior.

$$P(t) = \sum_{i=1}^{N} x_i(t)P_i(t) + P_{base}(t) \tag{1}$$

In this paper, NIALM is declared as a combinatorial optimization problem, which is formulated as the objective metric in Equation (2) and solved by a parallel computing accelerated GA.

$$err(t) = \text{argmin}\left|[P(t) - P_{base}(t)] - \sum_{i=1}^{N} x_i(t)[P_i(t) + \tau_i]\right| \tag{2}$$

where $\tau_i = c_i \cdot std(P_i(t))$.

In Equation (2), P_i (=mean($P_i(t)$)) obtained in advance involves profiling each of concerned electrical appliances based on historical data. That is, $P_i(t)$, in Equation (1), or P_i, in Equation (2), accounts for the real power that was absorbed by electrical appliance i and statistically computed for load disaggregation. P_{base}, in Equation (2), from $P_{base}(t)$, in Equation (1), is defined in a similar sense, and its standard deviation can be considered. In Equation (2), τ_i, a tolerance term for P_i, is considered, where std($P_i(t)$) is a function that returns the standard deviation of its input elements $P_i(t)$ from historical data and c_i is a constant that can be designed as a time-dependent parameter.

Figure 2 illustrates the principle of combinatorial search for load combinations with $x_i(t)$ in Equation (2) for load disaggregation, which aims to optimize load combinations by $x_i(t)$. In Figure 2, $P_i(t)$ represents the assumed power demand of the i-th electrical appliance. $x_i(t)$ associated with $P_i(t)$ is represented as a binary vector, evolved through metaheuristics, and used to minimize Equation (2) (the minimal error, $err(t)$, to be obtained for load disaggregation) between the summation of superimposed absorptions $\Sigma x_i(t)P_i(t)$ with an unmonitored base load $P_{base}(t)$ and the total load $P(t)$ ($P(t)$ to be approximated from a pack of $P_i(t)$ with $P_{base}(t)$). A metaheuristic algorithm, GA, is suitable for load disaggregation formulated in Equation (2) (the objective metric is the metric we are trying to optimize and the fitness metric is the algorithm's guide to doing so), where concerned electrical appliances are recognized by parallel computing accelerated GA for the declared objective metric. Concerned electrical appliances are constant or time-varying resistive, inductive, or capacitive loads.

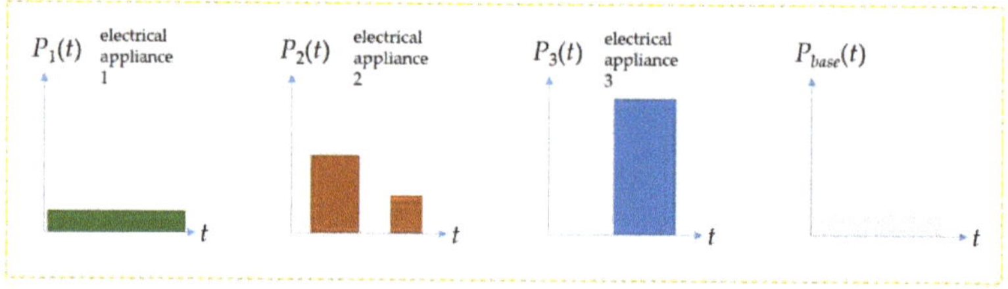

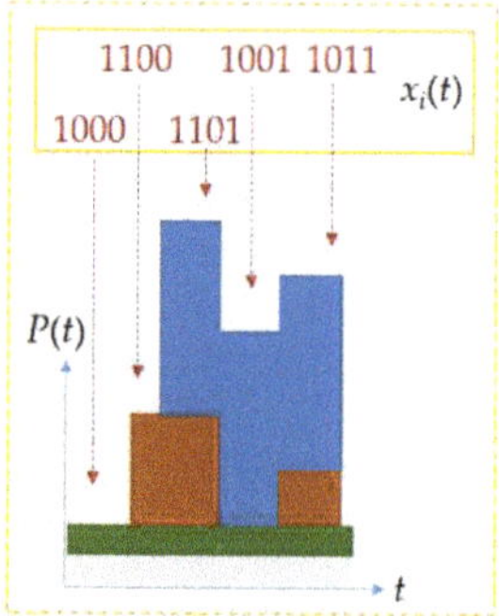

Figure 2. An illustration of the principle of combinatorial search for load combinations to $x_i(t)$ for load disaggregation in this paper, whereby load combinations by $x_i(t)$ are optimized. $x_i(t)$ is represented as a binary vector, evolved through metaheuristics, and used to minimize Equation (2) between the summation of superimposed absorptions $\Sigma x_i(t)P_i(t)$ with an unmonitored base load $P_{base}(t)$ and the total load $P(t)$.

The NIALM investigated in this paper has the main objective of recognizing concerned electrical appliances for (residential) DSM according to a composition of appliance-level real power consumption, which is disaggregated from total real power consumption acquired apart for load disaggregation. That is, with the minimum in Equation (2), acquired total real power consumption is the total of real power consumption by all individual electrical appliances concerned and operated where a base load should be considered. To Equation (2), a parallel computing accelerated GA as a load recognizer for load recognition of the NIALM in this paper is used to recognize the correct $x_i(t)$ to gain the minimum between the acquired total real power consumption and the sum of superimposed real power consumption by concerned electrical appliances. In this paper, real power consumption, P, is extracted, as the electrical feature for load recognition in Figure 1, from acquired total real power consumption, which is disaggregated into appliance-level real power consumption through the parallel computing accelerated GA-based NIALM of Equation (2). Concerned electrical appliances to be recognized can be constant or time-varying resistive, inductive, and capacitive loads [4,25].

2.1. GA-Based NIALM

GAs are a stochastic, population-based metaheuristic optimization (search) algorithm that can search for the global (quasi-)optimal solution metaheuristically to both constrained and unconstrained, NP-hard/NP-complete discrete or continuous optimization problems addressed, through (natural) selection operations that select current chromosomes for further gene recombination and through genetic operations involving (1) crossover operations that combine genetic information from selected parents' chromosomes to produce new offspring and (2) mutation operations that provide genetic diversity among population members. The evolutionary cycle is based on the principle of biological evolution. GAs that operate based on the principle of biological evolution have been proven to be an effective metaheuristics technique in many optimization problems.

In GAs, genetic operations, comprising crossover and mutation operations, and evolutionary operations, implementing the driving force of evolution (natural selection/Darwinism), are involved. See [26] for more details about a standard GA. In a GA addressing an optimization problem, a population of chromosomes as candidate solutions in a search space is evolved, from generation to generation, through genetic operations and evolutionary operations towards a global (quasi-)optimal solution. In an evolutionary cycle in a GA, a proportion of an existing population in its current generation is reproduced through selection operations and is used through crossover and mutation operations to breed a new population for the next generation. Selection operations select chromosomes based on a fitness-based selection procedure where chromosomes as candidate solutions are evaluated by a fitness function, and chromosomes with relatively high fitness are typically more likely to be selected. Also, crossover operations exchange subparts of two chromosomes, roughly mimicking the biological recombination between two haploid organisms. Finally, mutation operations randomly alter genes in some chromosomes, where an arbitrary bit of a chromosome is changed from its original state. In a GA, the population size depends on the nature of the problem addressed, but, typically, the population contains several hundreds or thousands of chromosomes/candidate solutions. Chromosomes are generated at random in the initial population. The evolutionary cycle is repeated, from generation to generation, until a termination condition such as a maximum of generations prespecified and reached has been met.

A GA used to address optimization problems is innately a parallel algorithm that can be run on a multicore processor. The workflow depicting the parallel computing accelerated GA used to solve Equation (2) for load recognition in the NIALM in this paper is shown in Figure 3. In the GA, function evaluations are farmed out to different processors on a multicore processor; they are executed in parallel. Figure 4 shows that $x_i(t)$ (at time t) in Equation (2) are encoded as a chromosome for an evolutionary cycle of the GA. To evaluate a chromosome with Equation (2), the GA decodes it as a computed summation of superimposed absorptions $\Sigma x_i(t)P_i(t)$. With an initial population of randomly generated chromosomes that are started in the parallel computing accelerated GA in Figure 3, successive generations are constructed through evolution. Fitter chromosomes are more likely to survive, based on selection, and to participate in genetic operations [27]. Here, (1) a roulette selection strategy choosing parents by simulating a roulette wheel in which the area of the section of the wheel corresponding to an individual is proportional to the individual's expectation associated with its scaled fitness value is conducted (a ranking method that scales raw fitness values based on the rank of each individual rather than its raw fitness value is used for fitness scaling), (2) a single-point crossover operator is used, (3) a bit-wise mutation operator is utilized, and (4) an elitist strategy guaranteeing that the solution quality does not decrease during evolution is also conducted.

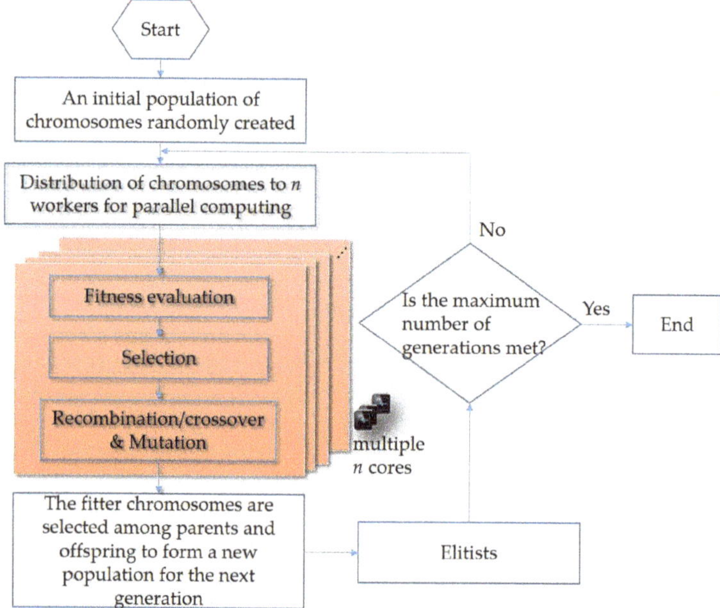

Figure 3. A workflow of the parallel computing accelerated GA used to solve Equation (2) for load recognition in the NIALM in this paper, where the evolutionary cycle is parallelized.

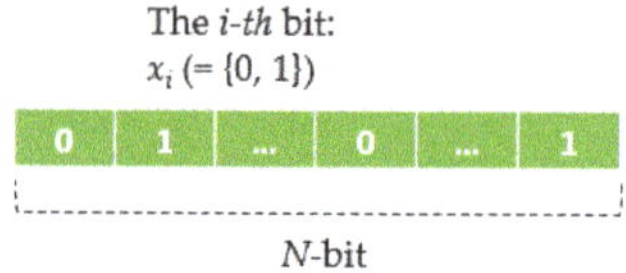

Figure 4. A chromosomal design encoding $x_i(t)$ (at time t) in Equation (2).

2.2. Feed-Forward, Multilayer ANN-Based NIALM

A feed-forward, multilayer ANN can be used to learn and distinguish from aggregated, extracted NIALM feature data for load disaggregation. In this paper, a comparative study where a feed-forward, multilayer ANN against the GA described in the previous subsection is used to address the same NIALM/load disaggregation problem, Equation (1), is conducted. As seen in Figure 5, a feed-forward, multilayer ANN contains an input layer, a number of intermediate hidden layers, and an output layer. The size of the input layer depends on the number of independent variables (extracted representative features) of considered feature data to be learned. The number of intermediate hidden layers with their size (the number of hidden neurons) specified in each hidden layer, can be determined experimentally through hyperparameter tuning where hyperparameters including the learning algorithm and the number of epochs for iterative rounds of learning affect how well the connectionism can represent the considered feature data (the hyperparameters are a set of parameters whose value is specified and used to control the learning process). The size of the output layer depends on the number of dependent variables (mutually exclusive classes) of considered feature data to be fitted.

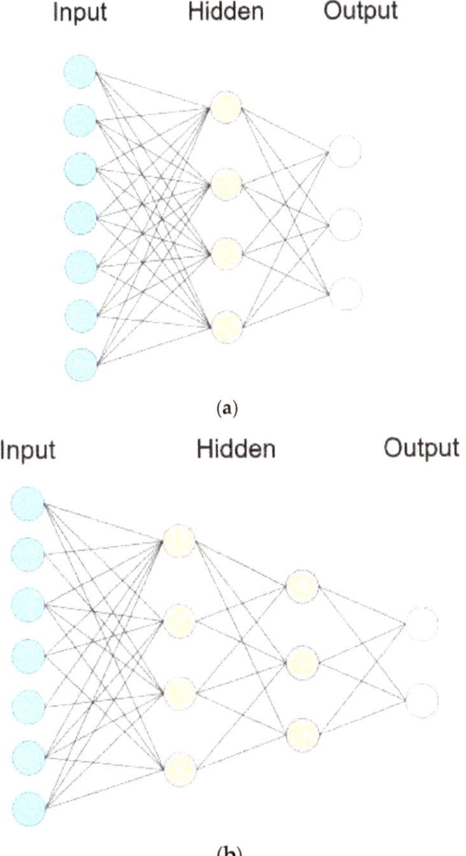

Figure 5. Representative ANNs, connectionisms, mimicking biologic NNs: (**a**) A connectionism is fully connected; (**b**) A connectionism considers dropout—one of the hyperparameters whose specified value is used to control the learning process of a connectionism—to prevent overfitting [28].

In this paper, F_1 score described in detail in the following subsection is conducted and used as the performance metric to indicate the effectiveness of the two parallel computing-accelerated AI approaches in load recognition for load disaggregation.

2.3. Performance Evaluation of Load Recognition by F_1 Score

In this paper, as shown in Equation (3), F_1 score [29] is used to evaluate the performance of the two parallel computing-accelerated AI approaches in load recognition for load disaggregation.

$$F_1 score = 2 \cdot \frac{Precision \cdot Recall}{Precision + Recall} \quad (3)$$

In Equation (3), precision is a ratio of the total number of correctly recognized positives to the total number of predicted positives. Recall (i.e., sensitivity or hit rate) is a ratio of the total number of correctly recognized positives to the total number of actual positives. See ref. [29] for more details about precision and recall. To summarize, F_1 score is the harmonic mean of precision and recall. A recognizer that produces no false positives has a precision of 1.0. A recognizer that produces no false negatives has a recall of 1.0. In Equation (3), an F_1 score reaches its best value at 1.0 (perfect precision and recall). The higher the score, the better the recognition performance. Besides F_1 score, we also use receiver operating characteristic (ROC) curves [29] to evaluate the performance of the two parallel computing-accelerated AI approaches in load recognition for load disaggregation. An ROC curve considering false positive rate (FPR) and true positive rate (TPR) is a graph showing the recognition performance of a recognizer examined at all recognition thresholds (or with different configuration settings) [29]. In an ROC curve, an error to a trained and validated recognizer can be computed by the Euclidean distance, from the perfect identification (FPR = 0, TPR = 1) til a resulting (FPR, TPR) [30,31] where FPR and TPR are defined as Equations (4) and (5), respectively. TPR is referred to as recall. As seen in Equations (4) and (5), ROC curves [29] are also used to evaluate the performance of the investigated methodology in load recognition, where TPR (a synonym for recall) defined in Equation (4) and FPR defined in Equation (5) are considered. See ref. [29] for more details about precision and recall.

$$TPR = \frac{TP}{TP + FN} \tag{4}$$

$$FPR = \frac{FP}{FP + TN} \tag{5}$$

In Equations (4) and (5), *TP* (true positives) means the data instances are recognized, and fit with reality. *TN* (true negatives) means the data instances are recognized, and the nonexistence fits with reality. *FP* (false positives) means the data instances are erroneously recognized as positives. Finally, *FN* (false negatives) means the data instances are incorrectly recognized as negatives.

3. Experimentation and Results

In this section, experiments are carried out to verify the validity of the investigated NIALM, by a publicly available UK-DALE (UK Domestic Appliance-Level Electricity) dataset [32]. The UK-DALE dataset contains records of power consumption measured and collected from five different households in the UK. In each house, the authors in [32] recorded both whole-house power consumption (power demand) from the mains every 6 s and power consumption by concerned individual electrical appliances every 6 s. Figure 6 shows the considered historical power demand on two typical days (Sunday 7 December 2014 (Figure 6a,b) and Thursday 4 December 2014 (Figure 6c,d)) in House 1 from the UK-DALE dataset, which is considered, parsed, and used to experimentally validate the performance of the investigated methodology in load recognition. The summarization of the UK-DALE dataset can be found in [32]. In Figure 6a,c, we show the total power demand in the mains. Also, we show the power demand by the concerned individual electrical appliances and all other submeters [32] considered together and treated as a single individual in power absorptions for load disaggregation. As shown in Figure 6b,d, the thin white gap between the power demand in the mains and the summed-up power demand illustrates the amount of power demand, base load, which is not concerned/metered. The concerned electrical appliances, including the considered submeters as a single individual in power absorptions to P_i in Equation (2), are listed in Table 1; their power demand is shown in Figure 6 and the base load, P_{base} in Equation (2), is assumed to have a constant value of 150.0 watts for simplicity's sake. Figure 7 shows the power demand of the several individual electrical appliances targeted in this paper and listed in Table 1. The behavior of the power demand by the electrical appliances can be found in [32]. For example, Figure 7a shows the power demand by the fridge running and doing its respective job of compressor on (state 1: its mean power consumption is ~90.0 watts with a standard deviation of ~44.0) or defrost (state 2: its mean power consumption is ~245.0 watts with a standard deviation of ~16.0). In this paper, the load classes considered from the targeted electrical appliances in Table 1 and recognized are shown in Table 2.

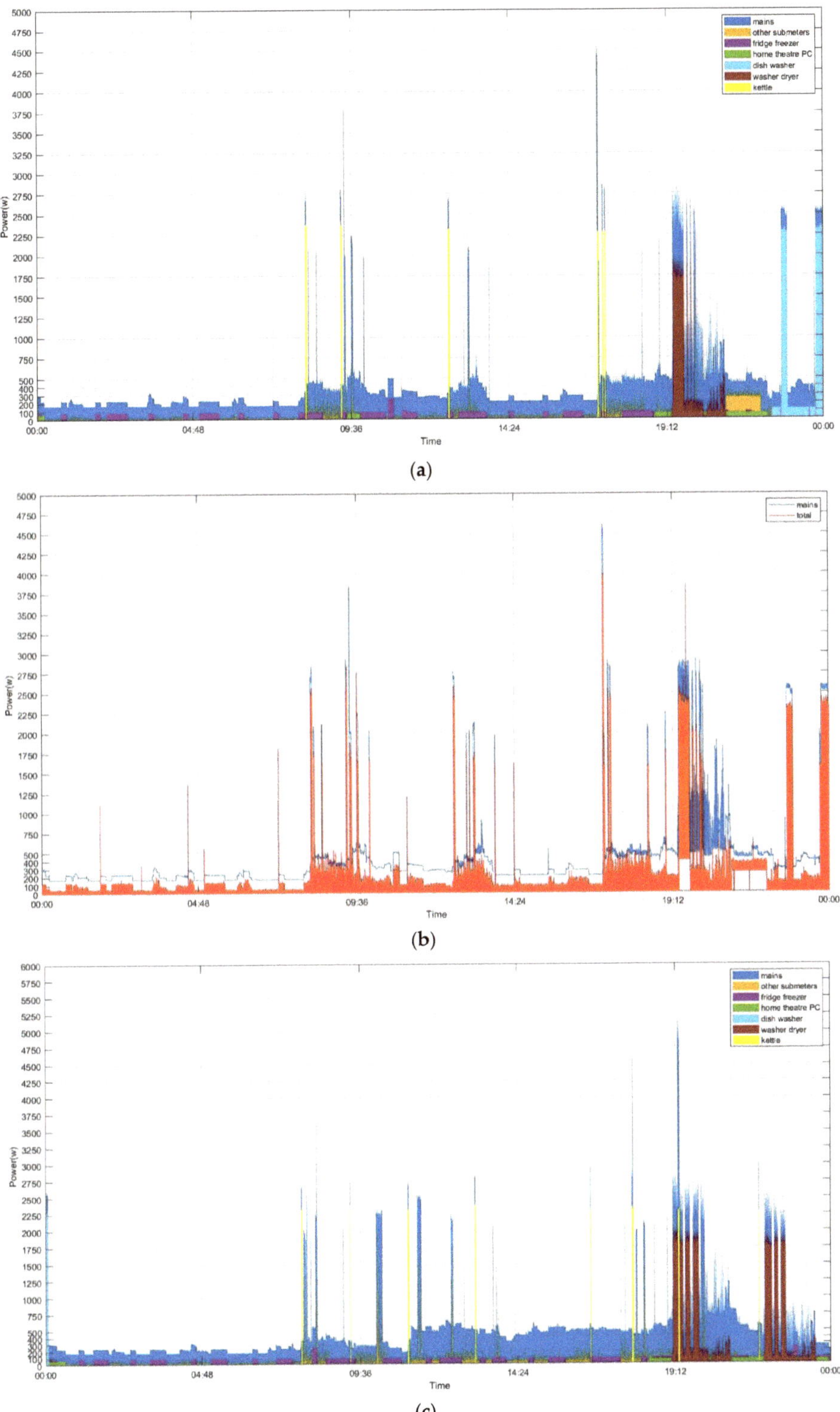

Figure 6. *Cont.*

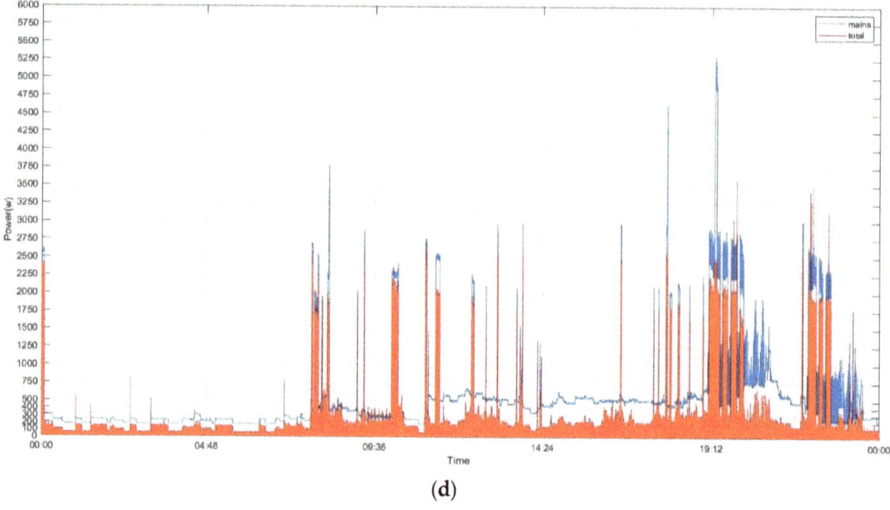

(d)

Figure 6. Considered historical power demand on two typical days in House 1 from the UK-DALE dataset parsed in this paper. The thin blue line in (**a**,**c**) shows the total (whole-house) power demand in the mains; the stacked, filled, and colored blocks show the power demand by the concerned individual electrical appliances and all other submeters [32] considered together and treated as a single individual in the power absorptions also concerned in load disaggregation. Base load in (**b**,**d**) exists in the house.

Table 1. Electrical appliances concerned and considered for load disaggregation in this paper.

Electrical Appliance	State 1		State 2 [3]		State 3		State 4	
	Mean [1]	Standard Deviation [2]	Mean	Standard Deviation	Mean	Standard Deviation	Mean	Standard Deviation
fridge	88.8	43.7	245.5	16.4	-	-	-	-
htpc (home theatre PC)	68.5	6.2	-	-	-	-	-	-
washer dryer	182.2	131.6	1833.1	152.9	-	-	-	-
dishwasher	116.0	15.1	2309.3	27.3	-	-	-	-
kettle	2323.7	132.6	-	-	-	-	-	-
other submeters	17.1	22.2	67.9	152.7	457.7	72.2	280.6	37.4

[1] P_i (= mean($P_i(t)$)): statistically computed, for the mean values, from the historical power demand data and stored in the database in Figure 1, where an eventless NIALM approach, the presented methodology, is shown; [2] τ_i: $c_i \cdot std(P_i(t))$; [3] In the GA, the multistate transitions from the same types of the concerned electrical appliances are mutually exclusive; illegal offspring are assigned an objective value of 1000 to Equation (2) (an illegal chromosome cannot be decoded to/as a solution; that is, such an illegal chromosome cannot be evaluated).

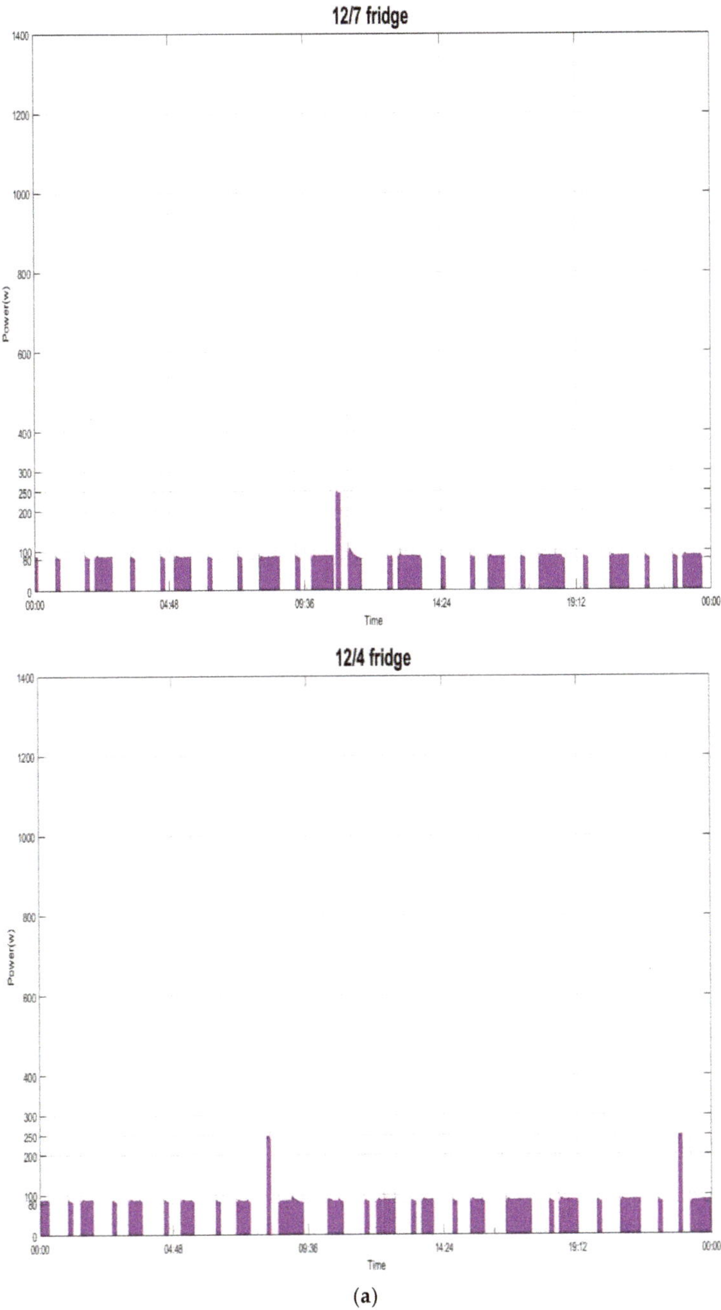

(a)

Figure 7. *Cont.*

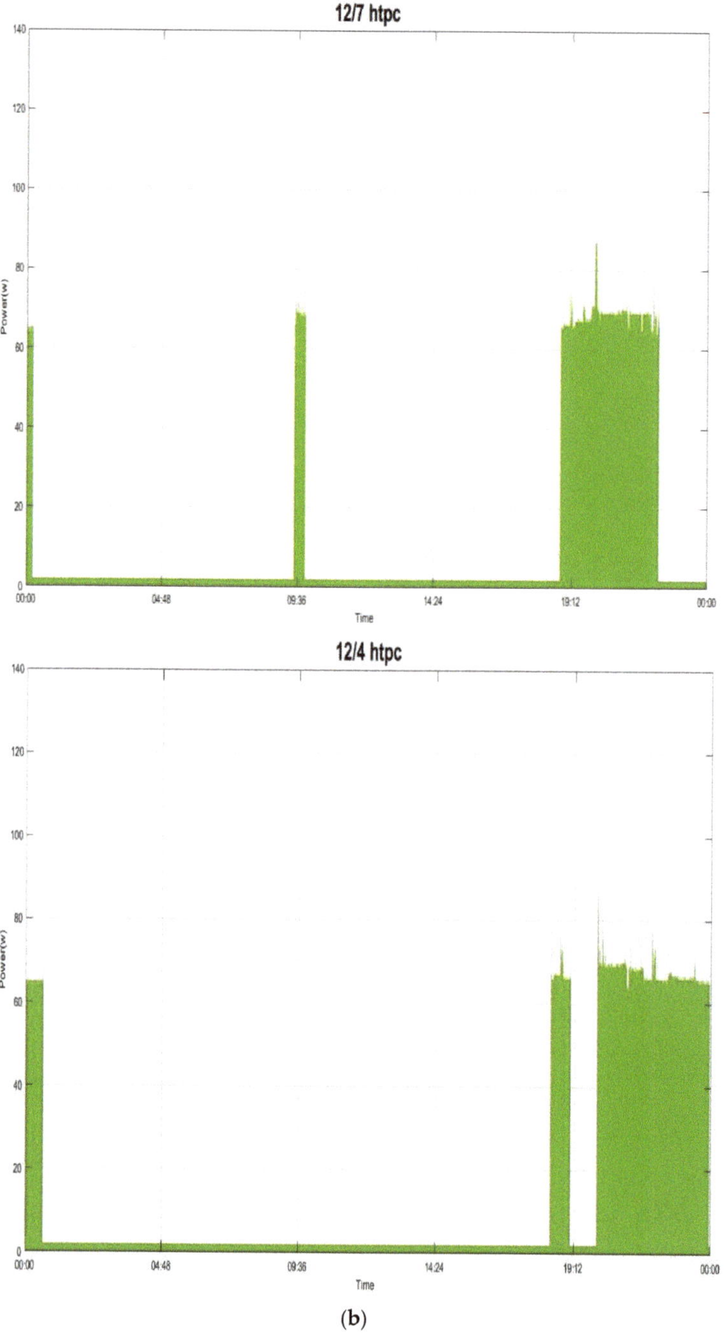

(b)

Figure 7. *Cont.*

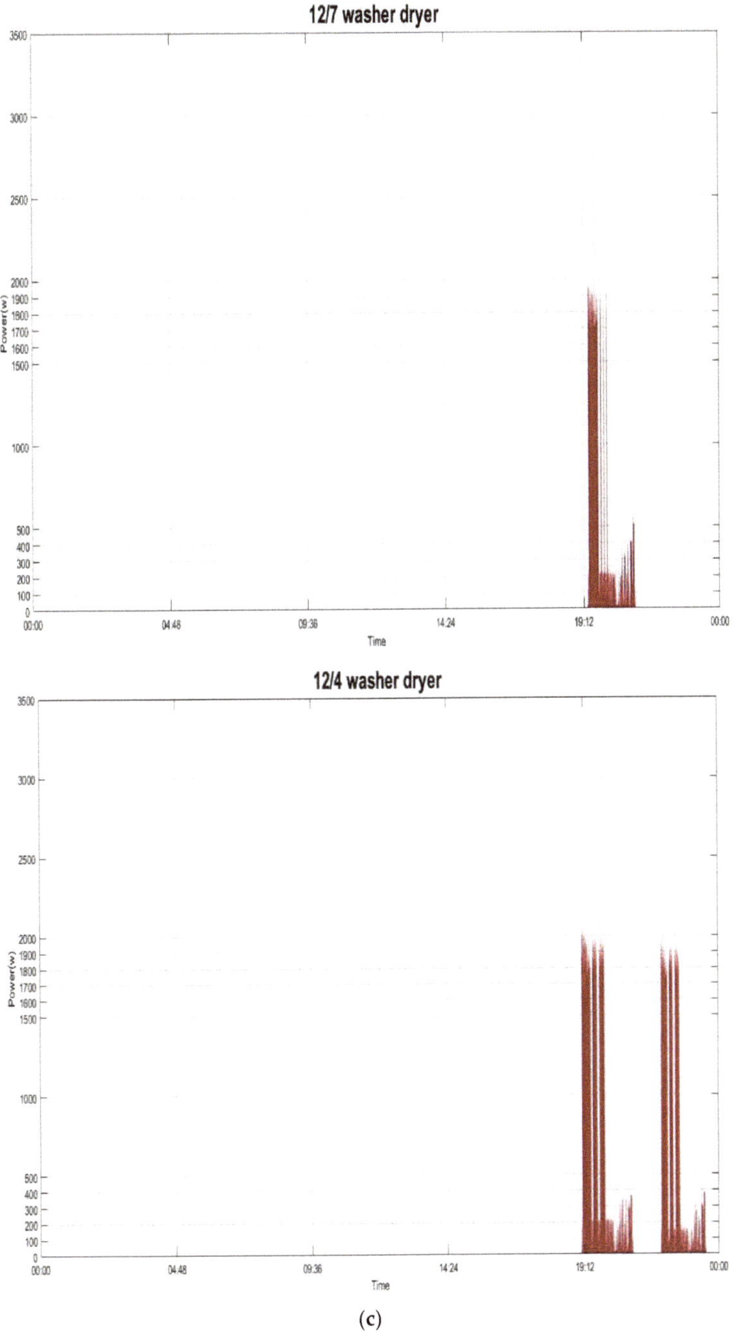

(c)

Figure 7. *Cont.*

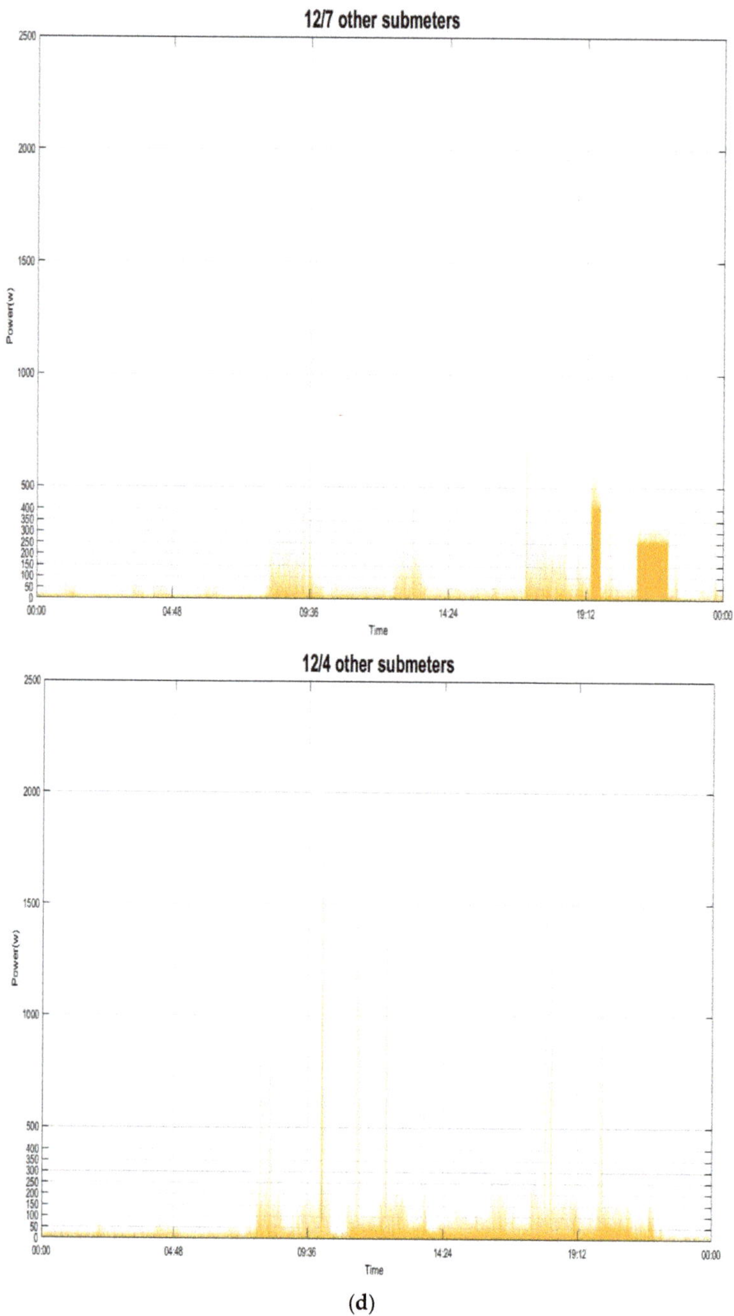

Figure 7. Shown power demand of most of the electrical appliances: (**a**) fridge; (**b**) htpc; (**c**) washer dryer; (**d**) other submeters.

Table 2. Twelve load classes considered from the targeted electrical appliances in Table 1 and recognized.

Class	Load
1	fridge, state 1
2	fridge, state 2
3	htpc, state 1
4	washer dryer, state 1
5	washer dryer, state 2
6	dishwasher, state 1
7	dishwasher, state 2
8	kettle, state 1
9	other submeters, state 1
10	other submeters, state 2
11	other submeters, state 3
12	other submeters, state 4

We used the UK-DALE dataset [32] as our reference dataset to experimentally validate the performance of the investigated methodology in load recognition; note that, in the dataset, data that were recorded from House 1 in the UK were considered in this experiment. In this experiment, a total of 4096 ($=2^{N(=12)}$) load combinations need to be recognized by the parallel computing accelerated GA in this paper (the total number of meters installed and used as ground truth in the house environment is 54 [32]), where a total of 208 composite power consumption (NIALM) data instances are disaggregated according to Equation (2). For each data instance acquired at time t and disaggregated, the parallel computing accelerated GA indicates the electrical appliances whose operation is active or inactive. In this experiment, the parallel computing accelerated GA is implemented in MATLAB® and run on an Acer Predator G3-710 Intel® Core™ i7-6700 CPU (3.40 GHz) (RAM: 16 GB) personal computer (PC), where for parallel computing the total number of available workers, n, on the machine is four. Note that running the parallel computing accelerated GA requires Global Optimization Toolbox™ [33] required with Parallel Computing Toolbox™ [34] for parallel computing. In this experiment, for simplicity, c_i in Equation (2) is set to 1.5, which can be determined through an exhaustive search for the house environment. Also, P_{base} assumed and estimated according to Figure 6b,d is 150.0 watts. Parameters for the parallel computing accelerated GA are specified below. The population size, *pop_size*, is 250. The initial population is created randomly in bit strings, where the total length of each chromosome is 12. The roulette selection strategy is used, and for this proportional selection procedure raw fitness values based on the rank of evaluated chromosomes, rather than their raw fitness value, are scaled. The single-point crossover operator is conducted; the crossover fraction of the population to be evolved is 0.55. The bit-wise mutation operator is used; the mutation rate is 0.01. An elitist strategy that guarantees a total of top ($0.05 \times$ *pop_size*) chromosomes to survive from their current population to the next population is also used. The maximum number of generations is 50. Finally, the fitness function is clarified as —E, where E is the declared objective function shown in Equation (2).

Figure 8 shows the evolutionary trajectory of the parallel computing accelerated GA to a disaggregated NIALM data instance. The resulting objective value obtained is 0.7. To the total 208 NIALM data instances where they are run over for load disaggregation, the parallel computing accelerated GA compared to a standard GA achieves, in terms of computation time, an acceleration of up to $3.49\times$ (=19.57 s/5.60 s).

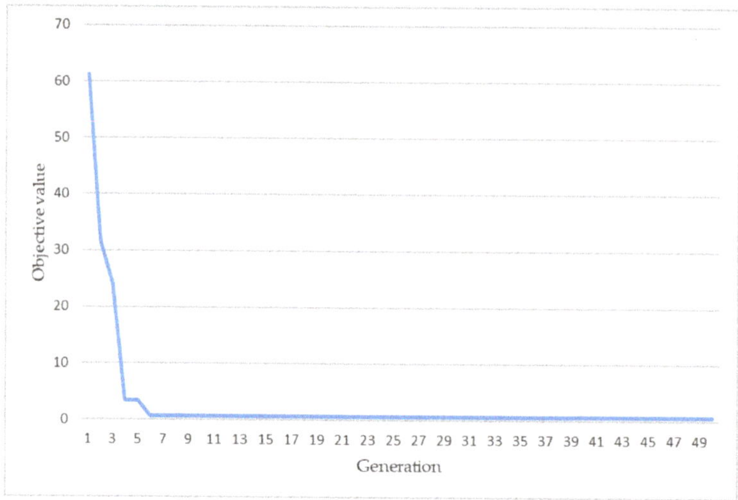

Figure 8. Obtained evolutionary trajectory by the parallel computing accelerated GA in this experiment.

In order to evaluate the performance of the parallel computing accelerated GA in load recognition, we examined the ROC curves, TPR and FPR in Equations (3) and (4), where their Area under ROC (AUC) is shown. Table 3 tabulates the load recognition results obtained by the parallel computing accelerated GA applied on the class-imbalanced data, where the F_1 score is shown to each load class. Table 4 also tabulates the load recognition results obtained by the parallel computing accelerated GA, where TPR vs. FPR is also shown to each load class. As shown in Table 4, the presented methodology got good load recognition results (against random guesses with AUCs of 0.5) for the fridge, kettle, and dishwasher, where activities of daily living (ADLs) [35] can be inferred from them for occupants in the house. The experimental results reported in this paper have shown the validity of the parallel computing accelerated GA-based NIALM for load disaggregation. In addition, the NIALM's achieved acceleration of up to 3.49× has been shown. The parallel computing accelerated GA can exploit parallel computing and, thus, reduce computation time. Computation time will increase exponentially when a large amount(s) of NIALM data is run over for load disaggregation (Equation (2)) in a large-scale evaluation of NIALM. Parallel computing will be exploited massively and the computation time will, thus, be reduced drastically. As shown in Tables 3 and 4, the performance of the presented methodology in load recognition needs a significant improvement, where the presented methodology suffers from similar P where the concerned electrical appliances or the load combinations of the concerned electrical appliances are identical. It will be improved. The improvement is shown below.

Table 3. Load recognition results obtained by the parallel computing accelerated GA.

Class	Precision	Recall	F_1 Score [1]	Number of Appliance Instances
1	0.36	0.81	0.50	58
2	0.00	0.00	0.00	25
3	0.39	0.47	0.42	88
4	0.05	0.04	0.04	26
5	0.00	0.00	0.00	68
6	0.57	0.14	0.22	87
7	0.12	0.36	0.18	11
8	0.14	1.00	0.24	4
9	0.48	0.57	0.52	97
10	0.28	0.29	0.29	51
11	0.00	0.00	0.00	7
12	0.25	0.14	0.18	35
Avg./total	0.32	0.33	0.29	557

[1] F_1 score is the harmonic mean of precision and recall, which is commonly used to evaluate a class-imbalanced problem addressed by a recognizer with a class-imbalanced dataset to be learned.

Table 4. Obtained TPRs vs. FPRs, for ROC curves, by the parallel computing accelerated GA, where for the 12 classes the AUCs obtained are also shown.

Class	FPR	TPR	AUC
1	0.56	0.81	0.63
2	0.11	0.00	0.45
3	0.54	0.47	0.46
4	0.11	0.04	0.46
5	0.15	0.00	0.43
6	0.07	0.14	0.53
7	0.15	0.36	0.61
8	0.12	1.00	0.94
9	0.53	0.57	0.52
10	0.24	0.29	0.53
11	0.10	0.00	0.45
12	0.09	0.14	0.53

The performance of the parallel computing accelerated GA in load recognition needs a significant improvement, although the algorithm is capable of recognizing the fridge/freezer, kettle, and dishwasher in the house environment for ADLs. A comparative study is conducted below, where a feed-forward, multilayer ANN as neurocomputing against evolutionary computing is used for the addressed NIALM problem. A network configuration of 1-15-12 of a feed-forward, multilayer ANN in Figure 5a is constructed, specified, and used, in this experiment, to address the same 208 NIALM data instances used before. The feed-forward, multilayer ANN was trained on 135 randomly sampled training data instances (~65% of the whole dataset) and tested on the remaining 73 test data instances. The training trajectory of the constructed, specified and used feed-forward, multilayer ANN is shown in Figure 9. The resulting mean squared error (MSE) is 0.055 (its initial MSE is 0.891). Table 5 tabulates the load recognition results obtained by the feed-forward, multilayer ANN, which has been well trained and validated on the class-imbalanced training data. In Table 5, the F_1 score is shown to each load class. Table 6 also tabulates the load recognition results obtained by the well-trained and -tested feed-forward, multilayer ANN applied on the class-imbalanced test data. In Table 6, TPR vs. FPR is also shown to each load class. The authors of [24] used the benchmark implementations, from NILMTK in [36], of the combinatorial optimization (CO) approach, which was developed in [1], as a load recognizer to perform load disaggregation for the UK-DALE dataset. A comparison among the CO, the parallel computing accelerated GA, and the feed-forward, multilayer ANN for load recognition is shown in Table 7. As shown in Tables 3–7, the feed-forward, multilayer ANN outperforms, in terms of load recognition, the parallel computing accelerated GA that is slightly superior in load recognition to the CO.

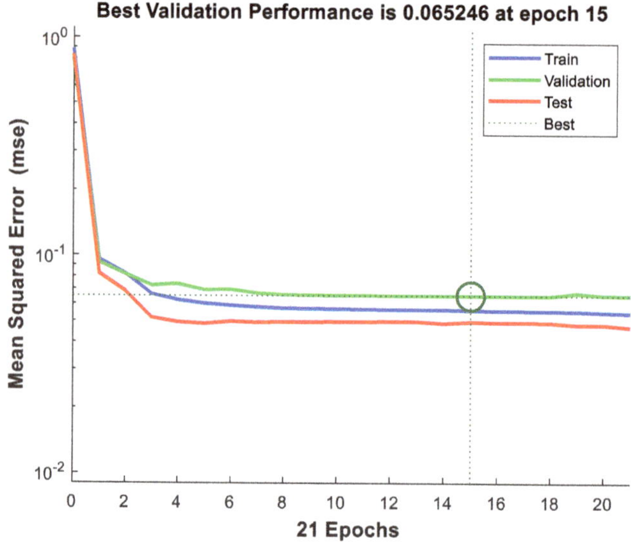

Figure 9. Training trajectory of the feed-forward, multilayer (1-15-12) ANN.

Table 5. Load recognition results obtained by the feed-forward ANN where it has been well trained and validated on the class-imbalanced training data.

Class	Precision	Recall	F_1 Score	Number of Appliance Instances
1	0.87	0.77	0.82	52
2	0.38	0.75	0.50	4
3	0.80	0.74	0.77	38
4	0.75	0.67	0.71	9
5	0.88	1.00	0.93	7
6	0.25	0.33	0.29	3
7	0.69	0.92	0.79	12
8	0.92	0.79	0.85	14
9	0.78	0.94	0.85	33
10	0.96	0.85	0.90	26
11	0.75	0.67	0.71	9
12	0.50	0.25	0.33	4
Avg./total	0.81	0.79	0.79	211

Table 6. Obtained TPRs vs. FPRs, for ROC curves, from the well trained and tested feed-forward, multilayer ANN applied on the class-imbalanced test data; for the 12 classes the AUCs obtained are also shown.

Class	FPR	TPR	AUC
1	0.29	0.77	0.74
2	0.07	0.75	0.84
3	0.20	0.74	0.77
4	0.03	0.67	0.82
5	0.02	1.00	0.99
6	0.04	0.33	0.65
7	0.08	0.92	0.92
8	0.02	0.79	0.88
9	0.23	0.94	0.86
10	0.02	0.85	0.91
11	0.03	0.67	0.82
12	0.01	0.25	0.62

Table 7. Comparison among the CO, the parallel computing accelerated GA and the feed-forward, multilayer ANN for load recognition.

Electrical Appliance	CO [1] [1]			The Presented GA-Based NIALM			The Presented ANN-Based NIALM		
	Precision	Recall	F_1 Score	Precision	Recall	F_1 Score	Precision	Recall	F_1 Score
fridge	0.30	0.41	0.35	0.18	0.41	0.25 [2]	0.63	0.76	0.68 [2]
dishwasher	0.06	0.67	0.11	0.35	0.25	0.2 [2]	0.47	0.63	0.54 [2]
kettle	0.23	0.46	0.31	0.14	1.00	0.24	0.92	0.79	0.85
Avg.	0.20	0.51	0.26	0.22	0.55	0.23	0.67	0.73	0.69

[1] Combinatorial optimization-based NIALM, which was implemented as a widget in the NILMTK developed for the purpose of performing preliminary work for NIALM data preparation and providing a few load recognition approaches for load disaggregation. [2] Across its all states.

In this experiment, more NIALM data instances, a total of 989 NIALM data instances, are sampled from the house environment and used to experimentally verify the performance of a feed-forward, multilayer ANN (Figure 5a) in terms of load recognition. A network configuration of 1-23-12 of a feed-forward, multilayer ANN was constructed, specified, and used. Also, the feed-forward, multilayer ANN was trained on 643 randomly sampled training data instances (~65% of the whole dataset) and tested on the remaining 346 test data instances. The training trajectory of the feed-forward, multilayer ANN is shown in Figure 10. The resulting MSE was 0.044 (the initial MSE was 1.110). The feed-forward, multilayer ANN can learn from the training data instances across all available CPU workers on the PC used. Table 8 tabulates the load recognition results obtained by the feed-forward, multilayer ANN where it has been well-trained and -validated. In Table 8, the F_1 score is shown to each load class. Table 9 also tabulates the load recognition results obtained by the well-trained and -validated feed-forward, multilayer ANN applied on class-imbalanced test data. In Table 9, TPR vs. FPR is also shown for each load class.

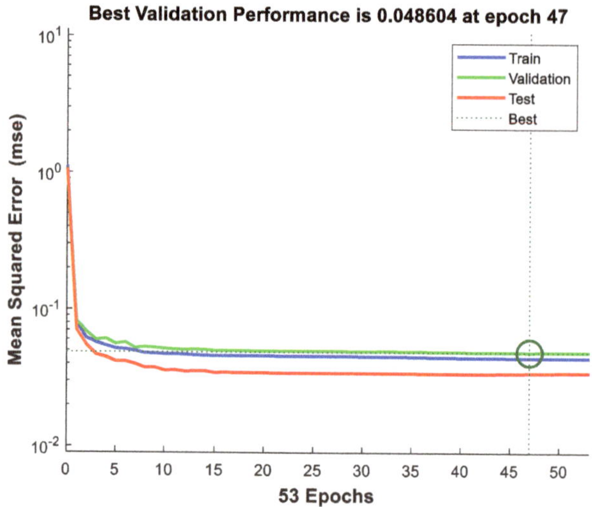

Figure 10. Training trajectory of the feed-forward, multilayer (1-23-12) ANN.

Table 8. Load recognition results obtained by the feed-forward ANN where it has been well trained and validated on the class-imbalanced training data.

Class	Precision	Recall	F_1 Score	Number of Appliance Instances
1	0.88	0.82	0.85	210
2	0.88	0.98	0.93	45
3	0.63	0.62	0.62	105
4	0.67	0.91	0.77	11
5	0.86	0.84	0.85	122
6	1.00	0.90	0.95	89
7	0.78	0.82	0.80	51
8	0.33	0.60	0.43	5
9	0.94	0.93	0.93	208
10	0.67	0.71	0.69	17
11	0.90	0.82	0.86	126
12	0.00	0.00	0.00	0
Avg./total	0.86	0.84	0.85	989

Table 9. Obtained TPRs vs. FPRs, for ROC curves, from the well trained and tested feed-forward, multilayer ANN applied on the class-imbalanced test data; for the 12 classes the AUCs obtained are also shown.

Class	FPR	TPR	AUC
1	0.17	0.82	0.83
2	0.02	0.98	0.98
3	0.16	0.62	0.73
4	0.01	0.91	0.95
5	0.08	0.84	0.88
6	0.00	0.90	0.95
7	0.04	0.82	0.89
8	0.02	0.60	0.79
9	0.10	0.93	0.92
10	0.02	0.71	0.84
11	0.05	0.82	0.88
12	0.02	nan	nan

As seen in Table 8, the 989 NIALM data instances make up a class-imbalanced dataset. Accuracy alone is not sufficient for evaluating a recognizer trained from class-imbalanced data, where in each class there may exist a significant disparity between positives (status: On) and negatives (status: Off). As a result, F_1 score, the harmonic mean of precision and recall, is conducted and used to address the class-imbalanced problem (in order to fully evaluate the performance of the feed-forward ANN in terms of load recognition). Various metrics in addition to F_1 score have been developed. Table 9 shows TPRs vs. FPRs, for ROC curves, obtained by the well trained and tested feed-forward, multilayer ANN applied on the class-imbalanced test data for NIALM, where one ROC curve can be shown per class (the maximum AUC is 1, which corresponds to a perfect recognizer as all positives above all negatives—100% sensitivity of no false negatives and 100% specificity of no false positives—are ranked). The authors of [24] used two different types of ANNs, deep NNs by autoencoder and long short-term memory (LSTM), as load recognizers to perform load disaggregation for the UK-DALE dataset. Comparison among the autoencoder, the LSTM and the presented feed-forward, multilayer ANN for load recognition is shown in Table 10. As shown in Tables 8–10, the feed-forward, multilayer ANN that gives similar performance, in load recognition, against autoencoder outperforms the LSTM. The load recognizer of the presented NIALM is a shallow neural network, not a deep neural work. As reported in this section, the presented feed-forward, multilayer ANN is able to discriminate the targeted electrical appliances from the house environment well.

Table 10. Comparison among the autoencoder, the LSTM and the presented feed-forward, multilayer ANN for load recognition.

Electrical Appliance	The Presented ANN-Based NIALM			A Deep ANN-Based Load Disaggregation by Autoencoder [24]			A Deep ANN-Based Load Disaggregation by LSTM [24]		
	Precision	Recall	F_1 Score	Precision	Recall	F_1 Score	Precision	Recall	F_1 Score
fridge	0.88	0.90	0.89 [1]	0.85	0.88	0.87	0.72	0.77	0.74
dishwasher	0.89	0.86	0.88 [1]	0.29	0.99	0.44	0.04	0.87	0.08
kettle	0.33	0.60	0.43	1.00	0.87	0.93	0.96	0.91	0.93
Avg.	0.70	0.79	0.73	0.71	0.91	0.75	0.57	0.85	0.58

[1] Across its all states.

4. Conclusions

A smart grid is a promising use-case of AIoT (AI across IoT) that enables bidirectional communication among utilities that come up with DR schemes for DSM and consumers who manage their power demands according to received DR signals. NIALM, a cost-effective load disaggregation approach for (residential) DSM, is able to disaggregate measured total power consumption into appliance-level power consumption based on unique electrical characteristics (features) extracted from electrical appliances concerned. In this paper, we have presented a parallel computing accelerated GA-based NIALM approach, where the presented methodology has been experimentally validated by the publicly available UK-DALE dataset as a reference. It is necessary to parallelize and thus accelerate metaheuristics by exploiting parallel computing, as metaheuristics such as GAs would require very high computational requirements due to its large amounts of data optimized, population-based candidate solutions evaluated as routines, and/or algorithmic iterations executed repeatedly. Besides the parallel computing accelerated GA-based NIALM approach, a feed-forward, multilayer ANN that can learn from training data instances across all available workers of a parallel pool on a machine in parallel computing addresses the same NIALM problem. Therefore, we performed a comparative study. Where, different load recognition approaches in the literature were compared. Comparison among the CO, the parallel computing accelerated GA and the feed-forward, multilayer ANN was shown. The feed-forward, multilayer ANN outperformed, in load recognition, the parallel computing accelerated GA that was slightly superior in load recognition to the CO. Moreover, a comparison among the autoencoder, LSTM, and feed-forward, multilayer ANN was shown. The feed-forward, multilayer ANN that gave similar performance, in load recognition, against the autoencoder outperformed the LSTM. As reported in this paper, the presented NIALM methodology whose performance in load recognition has been improved and compared is able to recognize the targeted/concerned electrical appliances from the house environment well for a future research direction of NIALM in ADLs. In this paper, AI speeded up in parallel computing has been developed and suited for NIALM. In the future, an additional electrical feature or more features will be considered for load disaggregation (Equation (2)). Moreover, high-performance distributed computing harnessing graphics processing units (GPUs) will be developed for the presented methodology in NIALM (batch load disaggregation) for its large-scale evaluation.

Author Contributions: Y.-C.H. conceived, designed, and performed the experiments as well as contributed related experimental tools/materials to analyze the experimental data. Y.-H.L. conceived, designed, and performed the experiments as well as contributed related experimental tools/materials to analyze the experimental data. Y.-H.L. also wrote the paper. C.-H.L. conceived, designed, and performed the experiments as well as contributed related experimental tools/materials to analyze the experimental data. All authors have read and agreed to the published version of the manuscript.

Funding: The Ministry of Science and Technology, Taiwan, under grant nos. MOST 109-3116-F-006-017-CC2 and MOST 109-2221-E-131-006-MY2 partly supported the work that has been done in this paper. First International Computer, Inc. (FIC), Taiwan, under the Industry-Academia Collaboration Project with grant no. O01109E048 partly supported the work as well.

Acknowledgments: The authors would like to sincerely thank the reviewers and editor for their valuable comments and suggestions on this paper.

Conflicts of Interest: The authors declare no conflict of interest.

References

1. Hart, G.W. Nonintrusive appliance load monitoring. *Proc. IEEE* **1992**, *80*, 1870–1891. [CrossRef]
2. Yu, J.; Gao, Y.; Wu, Y.; Jiao, D.; Su, C.; Wu, X. Non-intrusive load disaggregation by linear classifier group considering multi-feature integration. *Appl. Sci.* **2019**, *9*, 3558. [CrossRef]
3. Hosseini, S.S.; Agbossou, K.; Kelouwani, S.; Cardenas, A. Non-intrusive load monitoring through home energy management systems: A comprehensive review. *Renew. Sustain. Energy Rev.* **2017**, *79*, 1266–1274. [CrossRef]

4. He, H.; Lin, X.; Xiao, Y.; Qian, B.; Zhou, H. Optimal strategy to select load identification features by using a particle resampling algorithm. *Appl. Sci.* **2019**, *9*, 2622. [CrossRef]
5. Batra, N.; Singh, A.; Whitehouse, K. If you measure it, can you improve it? Exploring the value of energy disaggregation. In Proceedings of the 2nd ACM International Conference on Embedded Systems for Energy-Efficient Built Environments/ACM BuildSys'15, Seoul, Korea, 4–5 November 2015; pp. 191–200.
6. Froehlich, J.; Larson, E.; Gupta, S.; Cohn, G.; Reynolds, M.; Patel, S. Disaggregated end-use energy sensing for the smart grid. *IEEE Pervasive Comput.* **2011**, *10*, 28–39. [CrossRef]
7. Kong, X.; Zhu, S.; Huo, X.; Li, S.; Li, Y.; Zhang, S. A household energy efficiency index assessment method based on non-intrusive load monitoring data. *Appl. Sci.* **2020**, *10*, 3820. [CrossRef]
8. Massidda, L.; Marrocu, M.; Manca, S. Non-intrusive load disaggregation by convolutional neural network and multilabel classification. *Appl. Sci.* **2020**, *10*, 1454. [CrossRef]
9. Zhao, B.; Stankovic, L.; Stankovic, V. On a training-less solution for non-intrusive appliance load monitoring using graph signal processing. *IEEE Access* **2016**, *4*, 1784–1799. [CrossRef]
10. Guillén-García, E.L.; Morales-Velazquez, A.L.; Zorita-Lamadrid, O.; Duque-Perez, R.A.; Osornio-Rios, R.; de Romero-Troncoso, J. Identification of the electrical load by C-means from non-intrusive monitoring of electrical signals in non-residential buildings. *Int. J. Electr. Power Energy Syst.* **2019**, *104*, 21–28. [CrossRef]
11. Lin, Y.H.; Tsai, M.S. An advanced home energy management system facilitated by nonintrusive load monitoring with automated multiobjective power scheduling. *IEEE Trans. Smart Grid* **2015**, *6*, 1839–1851. [CrossRef]
12. Mueller, J.A.; Kimball, J.W. Accurate energy use estimation for nonintrusive load monitoring in systems of known devices. *IEEE Trans. Smart Grid* **2018**, *9*, 2797–2808. [CrossRef]
13. Kong, W.; Dong, Z.Y.; Ma, J.; Hill, D.J.; Zhao, J.; Luo, F. An extensible approach for non-intrusive load disaggregation with smart meter data. *IEEE Trans. Smart Grid* **2018**, *9*, 3362–3372. [CrossRef]
14. Lin, Y.H. Design and implementation of an IoT-oriented energy management system based on non-intrusive and self-organizing neuro-fuzzy classification as an electrical energy audit in smart homes. *Appl. Sci.* **2018**, *8*, 2337. [CrossRef]
15. Wu, X.; Gao, Y.; Jiao, D. Multi-label classification based on random forest algorithm for non-intrusive load monitoring system. *Processes* **2019**, *7*, 337. [CrossRef]
16. Lin, Y.H.; Hu, Y.C. Electrical energy management based on a hybrid artificial neural network-particle swarm optimization-integrated two-stage non-intrusive load monitoring process in smart homes. *Processes* **2018**, *6*, 236. [CrossRef]
17. Lin, Y.H.; Hu, Y.C. Residential consumer-centric demand-side management based on energy disaggregation-piloting constrained swarm intelligence: Towards edge computing. *Sensors* **2018**, *18*, 1365. [CrossRef]
18. Qi, B.; Liu, L.; Wu, X. Low-rate non-intrusive load disaggregation with graph shift quadratic form constraint. *Appl. Sci.* **2018**, *8*, 554. [CrossRef]
19. Zheng, Z.; Chen, H.; Luo, X. A supervised event-based non-intrusive load monitoring for non-linear appliances. *Sustainability* **2018**, *10*, 1001. [CrossRef]
20. Schirmer, P.A.; Mporas, I. Statistical and electrical features evaluation for electrical appliances energy disaggregation. *Sustainability* **2019**, *11*, 3222. [CrossRef]
21. De Baets, L.; Develder, C.; Dhaene, T.; Deschrijver, D. Detection of unidentified appliances in non-intrusive load monitoring using siamese neural networks. *Int. J. Electr. Power Energy Syst.* **2019**, *104*, 645–653. [CrossRef]
22. Fagiani, M.; Bonfigli, R.; Principi, E.; Squartini, S.; Mandolini, L. A non-intrusive load monitoring algorithm based on non-uniform sampling of power data and deep neural networks. *Energies* **2019**, *12*, 1371. [CrossRef]
23. Çavdar, İ.H.; Faryad, V. New design of a supervised energy disaggregation model based on the deep neural network for a smart grid. *Energies* **2019**, *12*, 1217. [CrossRef]
24. Kelly, J.; Knottenbelt, W. Neural NILM: Deep neural networks applied to energy disaggregation. In Proceedings of the 2nd ACM International Conference on Embedded Systems for Energy-Efficient Built Environments (ACM BuildSys'15), Seoul, Korea, 4–5 November 2015; pp. 55–64.
25. Yang, C.C.; Soh, C.S.; Yap, V.V. A systematic approach to on-off event detection and clustering analysis of non-intrusive appliance load monitoring. *Front. Energy* **2015**, *9*, 231–237. [CrossRef]

26. Lin, C.T.; Lee George, C.S. *Neural Fuzzy Systems: A Neuro-Fuzzy Synergism to Intelligent Systems*; International Edition; Prentice Hall (Pearson Education Taiwan Ltd.): Taipei, Taiwan, 2003; pp. 382–406.
27. Genetic Algorithm—MATLAB & Simulink3-MathWorks. Available online: https://www.mathworks.com/help/gads/genetic-algorithm.html; https://www.mathworks.com/help/gads/ga.html (accessed on 17 July 2020).
28. Hatcher, W.G.; Yu, W. A survey of deep learning: Platforms, applications and emerging research trends. *IEEE Access* **2018**, *6*, 24411–24432. [CrossRef]
29. Machine Learning Crash Course | Google Developers. Available online: https://developers.google.com/machine-learning/crash-course/classification/true-false-positive-negative; https://developers.google.com/machine-learning/crash-course/classification/precision-and-recall; https://developers.google.com/machine-learning/crash-course/classification/roc-and-auc (accessed on 29 April 2020).
30. Taveira, P.R.Z.; de Moraes, C.H.V.; Lambert-Torres, G. Non-intrusive identification of loads by random forest and fireworks optimization. *IEEE Access* **2020**, *8*, 75060–75072. [CrossRef]
31. Lu, M.; Li, Z. A hybrid event detection approach for non-intrusive load monitoring. *IEEE Trans. Smart Grid* **2020**, *11*, 528–540. [CrossRef]
32. Kelly, J.; Knottenbelt, W. The UK-DALE dataset, domestic appliance-level electricity demand and whole-house demand from five UK homes. *Sci. Data* **2015**, *2*, 150007. [CrossRef]
33. Global Optimization Toolbox—MATLAB-MathWorks. Available online: https://www.mathworks.com/products/global-optimization.html (accessed on 18 May 2020).
34. Parallel Computing Toolbox—MATLAB-MathWorks. Available online: https://www.mathworks.com/products/parallel-computing.html (accessed on 18 May 2020).
35. Devlin, M.A.; Hayes, B.P. Non-intrusive load monitoring and classification of activities of daily living using residential smart meter data. *IEEE Trans. Consum. Electron.* **2019**, *65*, 339–348. [CrossRef]
36. Batra, N.; Kelly, J.; Parson, O.; Dutta, H.; Knottenbelt, W.; Rogers, A.; Singh, A.; Srivastava, M. NILMTK: An open source toolkit for non-intrusive load monitoring. In Proceedings of the Fifth International Conference on Future Energy Systems (ACM e-Energy), Cambridge, UK, 11–13 June 2014; pp. 265–276. [CrossRef]

MDPI
St. Alban-Anlage 66
4052 Basel
Switzerland
Tel. +41 61 683 77 34
Fax +41 61 302 89 18
www.mdpi.com

Applied Sciences Editorial Office
E-mail: applsci@mdpi.com
www.mdpi.com/journal/applsci

www.ingramcontent.com/pod-product-compliance
Lightning Source LLC
LaVergne TN
LVHW071008170726
843515LV00006B/1111

* 9 7 8 3 0 3 6 5 6 5 0 8 8 *